Accidents
Will Happen

2/27/80

patience and consistence
work will pay.

Accidents Will Happen

The Case against Nuclear Power

by the

Environmental Action Foundation

Edited by

Lee Stephenson and George R. Zachar

With assistance from Gail Kovach

PERENNIAL LIBRARY
Harper & Row, Publishers
New York, Cambridge, Hagerstown,
Philadelphia, San Francisco, London,
Mexico City, São Paulo, Sydney

An earlier edition of this book was originally published under the title *Countdown to a Nuclear Moratorium*. Copyright 1976 by the Environmental Action Foundation.

FIRST EDITION

Designer: Stephanie Winkler

LIBRARY OF CONGRESS CATALOG CARD NUMBER: 79-2435

ISBN: 0-06-080505-6

79 80 81 82 83 10 9 8 7 6 5 4 3 2 1

Grateful acknowledgment is given for the following:
David Dinsmore Comey's article "The Browns Ferry Incident" originally appeared in *Not Man Apart*. Reprinted by permission of Friends of the Earth, Inc.

Lorna Salzman's article "The World's Most Dangerous Garbage." Portions of this article are excerpted from *Business and Society Review,* New York, N.Y., and *The Ecologist,* Wadebridge, Cornwall, England. Reprinted by permission of the publishers.

Kitty Tucker's article "A Little Radiation Goes a Long Way" originally appeared in *WIN* Magazine, November 9, 1978 (503 Atlantic Avenue, Brooklyn, N.Y. 11217). Reprinted by permission.

Charles Komanoff's article "Doing Without Nuclear Power" originally appeared in the May 17, 1979 issue of *The New York Review of Books.* Copyright © 1979 by NYREV, Inc. Reprinted by permission.

Richard Morgan's article "The Bargain Consumers Can't Afford" adapted from NUCLEAR POWER: THE BARGAIN WE CAN'T AFFORD. Copyright © 1977 by the Environmental Action Foundation, and from *The Power Line,* the May 1979 issue.

Dennis Hayes' article "Getting More From Less" reprinted by permission of Worldwatch Institute, Washington, D.C.

Helen Caldicott's article "The Ultimate in Preventive Medicine," adapted from NUCLEAR MADNESS: WHAT YOU CAN DO by Helen Caldicott, copyright © 1978 by Helen Caldicott. Reprinted by permission of Autumn Press, Brookline, Massachusetts.

Contents

What We Can't See *Is* Hurting Us

The Heart of the Issue: Money

The Solar Age Has Begun

It's Up To Us

ACKNOWLEDGMENTS

An early version of this book was published in 1972 under the title *The Case for a Nuclear Moratorium*. We would like to thank Sam Love and Larry Bogart for the original idea, and the Kaplan Fund for making that book possible. The book was expanded in 1976 by Richard Munson with the generous help of Marion and Warren Weber.

For coordinating the preparation of this anthology, we thank Claudia Comins. We appreciate the frequent advice of Gail Daneker and Richard Grossman of Environmentalists for Full Employment, Charles Komanoff and Elias Vlanton. Also very helpful were Randy Barber, Noreen Banks and Jeremy Rifkin of the People's Business Commission, Michael Bancroft, Thomas Cochran, Kate Donahue, Franklin Gage, Casey Grant, Linda Lotz, Jeremy Luban, Alden Meyer, James Ridgeway, Anne Seeley, Liz Waldo, and Annette Woolson.

We especially appreciate the contribution of articles and illustrations by many of those included in this book as a donation to Environmental Action Foundation.

Introduction

Ralph Nader

One of the questions I have asked leading supporters of
atomic energy is "If there were a major nuclear power
catastrophe in this country, what would be the future of
this technology?" Senator John Pastore and Dr. Alvin
Weinberg (formerly director of the Oak Ridge National
Laboratory) replied that politically that would be the end
of the industry. "The whole kit and caboodle would go,"
said Pastore who was the leading promoter of nuclear
power in the Senate for many years.

Their replies emphasize the fundamental uncertainty
and unreliability that characterize the pockmarking of our
country with hundreds of nuclear plants, thousands of
yearly movements of vehicles transporting radioactive
materials, dozens of radioactive waste dumps, many piles
of uranium tailings near uranium mines emitting radon gas
over huge land areas, and several reprocessing plants. Generating electricity by means of such a complex, vulnerable
technology threatens the economy and puts us on a reckless future course. No other form of energy, in the event
of a major catastrophe, would be so susceptible to public
repudiation.

But, then, what other form of energy is so unsafe that it
is uninsurable and requires a law (the Price-Anderson Act)
that limits liability and prevents the injured, sick, and dam-

aged from recovering compensation to the full limit of the assets of the utility or reactor manufacturers? What other form of energy produces wastes that are lethal for some 250,000 years and for which there is no safe, proven storage technology? What other form of energy can produce a disaster rendering hundreds of square miles, or more, of this country uninhabitable? The millions of people in the Chicago, Boston, New York, and San Diego areas—all particularly close to nuclear plants—face an extraordinary threat to their lives and health. What other form of energy is more likely to invite future generations—afflicted with cancer, birth defects, and unusable natural resources—to curse the generation whose unthinking corporate and political leaders unleashed the electric atom on an unsuspecting, uninformed public?

There are more questions that citizens must ask of their electric utility and their government. What other form of energy is so vulnerable to sabotage, theft, and other disruptions—all requiring, of course, expensive security measures that can lead to invasion of privacy and denial of civil liberties to innocent people? What other form of energy is associated by law with a population evacuation plan in case of accident? No coal-fired, oil-fired, gas-fired, geothermal, or solar energy system needs preparedness plans for millions of people to flee their homes, workplaces, and schools within a radius of up to 100 miles from the facility.

Clearly, as this volume of reports on the ghastly industrial folly known as nuclear power shows, you cannot avoid the subject without being subjected to its costly and hazardous consequences. Because all Americans are at risk from nuclear power, they should share part of the solution. To consumers, workers, taxpayers, and parents, the stakes are real—as the people living near Three Mile Island, with its near catastrophe, have discovered.

This book describes the solutions: ways homeowners and consumers can eliminate any need for nuclear power, and policy recommendations that citizens can demand be implemented. If this readable volume can be summarized in one sentence, it is that atomic energy is too expensive, too unsafe, too unreliable, and unnecessary.

What mired our society in this technological Vietnam was a combination of governmental and industrial secrecy coupled with the cost-plus investment structure of the electric utilities. For years, the facts, doubts, failures, and risks of nuclear power were matters of official secrecy—not to be disseminated or discussed with the public. I recall in the early 1960s questioning nuclear scientists about what could go wrong with a reactor. They would, uniformly, dismiss any concerns with phrases like "negligible" or "vanishingly small." Refusing to concede the possible, they had become prisoners of a technocratic faith that made them believe that there were imminent solutions to transcendent problems of nuclear technology.

Secrecy garnished by bland reassurances delayed the arrival of the first books cautioning people about atomic energy until the late 1960s. After the breakthrough, a small number of scientists, lawyers, and citizens began breaking down the walls of secrecy surrounding the nuclear-power issue. The media became more and more interested. A veritable flood of disclosures and leaks ensued, variously associated with freedom-of-information requests, lawsuits, congressional inquiries, and resignations under protest from government and industry.

Information being the currency of democracy, it is not surprising to find a growing democratic awareness of the hazards and costs of nuclear energy. But its cost-plus structure continues to shield this energy source from market disciplines that long ago would have stopped this complex and dangerous way of boiling water. As long as the utilities earn more money by investing more in expensive technology, and as long as the government subsidizes nuclear technology directly and—by the Price-Anderson Act—indirectly, the incentive for the utilities to reevaluate their reactor programs remains small. The policy of these monopolies is to make consumers pay for nuclear accidents (as at Three Mile Island) and for costly reactor-plant shutdowns due to repairs or retrofits—regardless of the extent of negligence involved.

By way of contrast, the reactor manufacturers cannot so easily pass on their losses. It is known that General Electric wants to get out of the reactor business. The corporation privately informed the Department of Energy of its intentions in 1978 and was asked not to close its reactor division because it would produce a domino effect on the rest of the industry. General Electric's reason was simple—it was losing money because it was not getting orders. In the past four years, the utilities have canceled or postponed over 140 nuclear-plant orders. Other reactor manufacturers are in much the same position as General Electric: one high executive told me he couldn't wait for his company to close down their reactor division.

The expansive dreams of nuclear power proponents are no more. Just a few years ago, they were predicting that by the year 2000 over a thousand nuclear plants would be operating. In California alone, one nuclear plant for every 15 miles on the average along the Pacific Coast was projected, unless they were clustered in nuclear parks. Now, only cost-plus and government subsidy make any strain of that dream a continuing possibility for the proponents. And only one more catastrophe need occur before the nightmare takes over. Nuclear Regulatory Commission Chairman Joseph Hendrie stated in April 1979 that the industry could not endure one more Three Mile Island—and that accident was merely a *near* catastrophe.

In the coming months, there will be ever stronger public awareness that nuclear power can be replaced by energy conservation, advanced ways of using conventional fuels, and the arrival of increasingly economical solar technologies. Clearly, a society that wastes at least 50 percent of its energy can replace the less than 4 percent of its energy that is at present nuclear. The myth of its necessity is the final one to be dissipated by the advocates of a nonradioactive future for America.

Ultimately, the struggle over nuclear power will be decided in the political arena at the local, state, and national levels. This book by the Environmental Action Foundation should make a signal contribution to the expansion of citizen awareness and constructive action for an abundant, efficient, and benign energy supply.

It Was Too Good
to Be True

_____

Sam Love

The nuclear industry's present problems with accidents and public protests are a far cry from the fantasies once envisioned for atomic energy. The most zealous boosters of nuclear power aggressively promoted the idea that nuclear energy would, more than any other technology, create a wonderful technological utopia where nature would be controlled, work would be virtually eliminated, and unlimited, cheap power would be at everyone's fingertips. America's, indeed the world's, wildest fantasies would be realized.

Now the only thing that appears to be unlimited about nuclear power is the cost of our electric bills. But not too long ago the future looked very, very different.

Soon after the fissioning of uranium was announced, the popular press and some scientists and politicians seized the unexplored potential of nuclear power as the ultimate technical panacea—almost a godsend to a depression-weary public. Articles extolling the virtues and promise of nuclear energy filled newspapers and magazines. Now, after the public has seen the effects of atomic attacks on

Japan, the spread of nuclear weapons, and the mounting crises surrounding atomic power plants, these stories take on an almost farcical tone.

In a 1941 issue of *Popular Mechanics,* Dr. R. M. Langer of the California Institute of Technology predicted the development of "power plants the size of a typewriter." Using a single pound of uranium, he calculated, such a power source would provide the energy of 250,000 gallons of gasoline. A car powered by this uranium load could go 5 million miles without refueling.

Social benefits would abound, he predicted. "We can look forward to universal comfort, practically free transportation and unlimited supplies of materials." Cost would be no problem, Langer speculated, because these reactors would provide cheap power in every home and industrial plant, at a cost of "less than one-tenth of a cent per kilowatt-hour." (The national average for nuclear plants is now more than 15 times his projected cost.)

Langer felt that typewriter-sized nuclear power plants would be ideal for hobbyists. "There is no end to the practical applications that amateurs can work out once the energy source is available," he wrote. Among other uses, he foresaw machines that would use this new power source to create intense heat to weld our highways together by fusing dirt into pavement.

Langer wasn't alone in his fantasies. A president of the Society of Automotive Engineers in the early 1940s predicted atomic auto engines as small as a person's fist. A book called *Atomic Energy in the Coming Era,* by David Dietz, the science editor of the Scripps-Howard newspaper chain, described "artificial suns" made from chunks of uranium mounted on towers. With these we could finally bring under control one of humanity's arch-nemeses—the weather. Nature wouldn't stand a chance once we had nuclear energy:

No baseball game would be called off on account of rain in the Era of Atomic Energy. No airplane would bypass an airport because of fog. No city would experience a winter traffic jam because of heavy snow. Summer resorts would be able to guarantee the weather and artificial suns would make it as easy to grow corn and potatoes indoors as on the farm.

U.S. government officials were among the atom's most committed early promoters, and their predictions among the most far-fetched. In retrospect, it seems that some of these officials did a grave disservice by distorting the facts about nuclear energy.

Harold Stassen, who in 1955 was a special assistant to President Eisenhower on disarmament, wrote an article for the *Ladies Home Journal* that appeared under the headline: "Harold Stassen, the man President Eisenhower has chosen to help the world end war, tells our readers how we can use atoms for peace."

Just think, it is estimated that one pound of raw atomic fuel—uranium-235 or plutonium—could generate as much power as 1500 *tons* of coal. . . . To think about it all gets into the realm of the really fantastic.

You hear talk from electronic engineers about houses without any wiring, but with portable, cheap baby watchers; bedmakers; cordless irons, lamps, and toasters; and light, thin, motorless TV sets, refrigerators, and air conditioners. Visionaries in the housing field dream of roll-up walls, bubble-shape houses that keep themselves clean, and prefabrication methods that make it possible to "move" from an old house to a new one without leaving your garden and your neighbors.

Yes, these things are fantastic—and yet they are not. With low-cost, limitless power—and that's what atomic energy is potentially—all sorts of dreams can come true.

Stassen reassured the public that nuclear power promised to be far less harmful to humanity than coal and oil fuels.

> The reactor is not, as some nervous people seem to fear, a pint-sized bomb that is about to explode. Its lead clothing is so thick that you can walk right up to it—as I have—and place your hand on its side and pat it as gently as you would a baby, and no increase of radiation will register on your "dosimeter."

Lewis L. Strauss, chairman of the Atomic Energy Commission under the Eisenhower administration, in a 1955 interview in *Reader's Digest*, invoked a higher authority for authorization to proceed with nuclear development.

> We are living in an era that seems designed to test the courage and faith of free men. Yet I do not believe that any great discovery of the atom's magnitude came from man's intelligence alone. A higher intelligence decided that man was ready to receive it.

Strauss was also the source of the now-famous claim that electricity from nuclear power would be "too cheap to meter."

Even President Truman got in his licks as a promoter of the atom. In his State of the Union message to Congress in 1950 he predicted:

> In the peaceful development of atomic energy, we stand on the threshold of new wonders. The first experimental machines for producing useful power from atomic energy are now under construction. We have made only the first beginnings in this field, but in the perspective of history they may loom larger than the first airplane, or even the first tools that started man on the road to civilization.

David Lilienthal, the first chairman of the Atomic Energy Commission, which was set up in 1946 to carry on all atomic work for both military and peaceful uses, was another effusive promoter of nuclear power. He wrote in a 1949 issue of *Collier's:*

> From now on we can travel at the speed of enlightenment or we can dawdle and fool and politick around, and postpone or lose the greatest opportunities man has known. . . .
>
> It is a great privilege to be alive at a time in the world's history when a discovery akin to finding fire or electricity comes along. The sooner men accept that fact and are stimulated by it, the sooner in my opinion will we enjoy the now incredible possibilities of atomic science. . . .
>
> There used to be three Rs that everyone had to learn: reading, 'riting, and 'rithmetic. To this we must now add a fourth R: radiation.

In an interview published that same year in *U.S. News and World Report,* Lilienthal was asked if atomic energy might conceivably be a means of prolonging human life and producing more food for the world.

> I think it is very difficult to overstate the effects in just those terms. . . . I would say within a decade, if we have made headway on our chief problem—the problem of peace—we can look back on this decade as having a profound effect on health and the lengthening of life.
>
> I should think such a decade could mean at least as much as a century of some earlier period in the development of longevity and health, and in improvement in food production.

Fourteen years later Lilienthal looked back on this era and criticized himself and others for such unequivocal promotion of atomic energy. In 1963, in his book *Change, Hope, and the Bomb,* Lilienthal reflected:

> The basic cause [of the unquestioning early optimism], I think, was a conviction, and one that I shared fully and tried to inculcate in others, that somehow or other the discovery that had produced so terrible a weapon simply *had* to have an important peaceful use. . . . We were grimly determined to prove that this discovery was not just a weapon. This lead perhaps to wishful thinking, a wishful elevation of the "sunny side" of the atom.

Dr. Ralph E. Lapp, still a nuclear advocate today, presented "Fifty Little-Known Facts About the Atom" in a 1952 issue of *Collier's.* He was then a 34-year-old nuclear physicist and veteran of the Manhattan A-bomb project.

> Q: When atomic power plants are perfected, will it be safe to have them located inside city limits?
>
> A: Yes. Atomic power generating stations will be made even more reliable than steam plants are today. All plants will be equipped with elaborate automatic control devices to prevent the atomic reaction from running wild and blowing up.
>
> Q: Do workers in our atomic plants have normal offspring?
>
> A: Yes. Children born in Oak Ridge [Tennessee] and Los Alamos, New Mexico are perfectly normal. If anything, as one doctor remarked, judging by the baby carriage parade at Oak Ridge, there seems to be remarkable fertility among people who work in atomic plants.

Other industries found their own angles from which to fantasize. Dr. Harry L. Fisher, then president of the American Chemical Society, wrote in a 1954 issue of the *Ameri-*

can Magazine about the prospects for atomic energy and
an atomic battery that he said would produce electricity
directly from radioactive substances without the need for
shielding.

> The first battery is the size of a thimble and produces
> about as much power as a fly in flight. It is the barest
> beginning. Still, it makes conceivable not only the
> atomic auto, but such other wonders as home washing
> machines with their own built-in atomic power source,
> and even tiny wristwatch radios. . . .
>
> Atomic power will transform the appearance of
> your home town. If you live in a community darkened
> ened by grime and afflicted with smog from power
> plant or factory smokestacks, you can look forward to
> seeing your town transformed into a clean, healthful
> place. Atomic furnaces, unlike coal furnaces, need no
> smokestacks. . . .
>
> Many parts of the American West and Southwest are
> now considered wastelands. Yet the land is rich and
> fertile. It can be made fabulously productive by water
> pumped there by cheap atomic power, as can many
> areas in North Africa, the Near East, Mexico, and Aus-
> tralia.

The electric-utility industry did not jump on the nuclear
promotion bandwagon until well into the 1950s—after
Washington put the atomic industry in a "can't lose" position by providing subsidies, tax shelters, and protection
against insurance claims for nuclear accidents. When
America's utilities, besieged by reactor sellers offering loss-
leader prices and guaranteed a profit on nuclear purchases
through cost-plus utility rates, started to "go nuclear," they
too joined the promotional orgy.

The predictions and promises of the early promoters of
atomic energy were brought to the public's attention with

Tony Auth. © 1979, The Philadelphia Inquirer. The Washington Post Writers Group.

the help of one of the most effective governmental propaganda programs ever created.

Beginning in the early 1950s, the Atomic Energy Commission spent hundreds of thousands, and sometimes millions, of dollars every year on films and other promotional materials. At its peak in 1972, the AEC film catalogue contained 232 films, among them such cinematic puff pieces as *Go Fission, Opportunities Unlimited: Friendly Atoms in Industry, Atom and Eve, The Alchemists' Dream,* and *The Day Tomorrow Began.* Many were aimed at student and nonscientific audiences. They were available for free loan and stressed that our future was inextricably tied to "safe, cheap" nuclear power.

Large mobile exhibit halls toured shopping centers and the state-fair circuit. Smaller displays on the "Peaceful Atom" were offered free to science museums.

These exhibits used elaborate display techniques to make their point. For example, a visitor entered the main display area of the AEC-funded Museum of Atomic Energy in Oak Ridge, founded in the 1940s and now renamed the Museum of Science and Energy, by walking down a narrow hall that displayed what life was like before electricity. At the end of the hall the visitor turned a corner and was suddenly confronted with three large futuristic white spheres with viewing screens. These screens showed changing images of "the atom at work" and broadcast a taped message assuring listeners of the safety of nuclear power. After this, visitors strolled into a simulated breeder reactor core for a pitch about that proposed new generation of reactors—without a word about the special dangers that have hindered its development. Next stop, a giant mockup of the interior of an even more distant and theoretical kind of nuclear energy—a fusion reactor, complete with bluish strobe lights.

Atomic-energy promotion is now under the control of the Department of Energy (DOE), which replaced the AEC. Promotional material prepared for DOE is now much more balanced, allowing other energy alternatives to compete with atomic energy for the public's attention. However, pro-nuclear prejudice in the government remains. Nuclear energy still receives a far larger portion of the DOE budget than other forms of energy. And in 1976, DOE tried to help discourage Californians from voting for a state referendum that would have placed restrictions on the construction of nuclear power plants. The agency sent hundreds of thousands of free pro-nuclear promotional brochures to utilities for use in fighting the referendum. The referendum failed.

There have been responsible energy experts who have tried to present a more realistic picture in the media, but it must have been hard to compete with the grand predictions coming from the promoters. Louis Cassels, in a *Harper's* article following Truman's 1950 State of the

Union address quoted earlier, tried to bring readers down to Earth:

> As President Truman's message to Congress amply demonstrates, the dream of atomic power for peacetime use has been held before us nearly as long and almost as vividly as the nightmare of atomic war.
>
> Indeed, in political oratory and some segments of the press, the issue has been so sharply drawn and the alternatives so luridly painted that one might be led to believe the threat of war is the only roadblock between us and atomic millennium. Let peace be assured, and great cities will rise in the desert. Atomic engines will do all the work, and men will be free to tour the country in the Nuclear-8 sedans.

The scientists and engineers who have actually begun to develop nuclear energy, Cassels wrote, "see clearly how long and rugged is the rest of the way, and they know that keeping the world at peace is only one—albeit the biggest —hurdle ahead."

Bill Perkins, a staff member of the Atomic Industrial Forum, the nuclear power industry's trade association, warned the International Science Fiction Convention in 1977 that spreading wild fantasies about nuclear power was actually counterproductive because it raised "expectations of the new technology to unrealistically high levels." He cautioned science writers against "racing into print with what the public wanted to read."

So maybe the desire for technological salvation is the real culprit. Wouldn't it be wonderful if we never had to miss the golf game because of rain? Or never pay another utility bill? Or never have to stop at the service station again? Or never grow old? Or . . .

The real world makes each of us pay a price for being

here. As both ecologists and economists love to say, there
is no such thing as a free lunch.

But that lesson is a tough one. And it is far too negative
to excite people, so we still pursue the fantasies. In spite of
considerable opposition, research and development continues on the nuclear breeder—the reactor the government and nuclear industry promise us will create more fuel
than it consumes.

And if that doesn't work, or plutonium—a by-product of
both conventional and breeder reactors and one of the
most poisonous elements known—is too difficult to handle,
we can always place our faith in fusion, the process that
would imitate the nuclear reaction in the sun. Optimists
now predict that we can get all the energy we can use from
fusion using the deuterium in ocean water. "In fact, for the
energy you could get out of it, deuterium from the oceans
would be only one-hundredth as expensive as coal," claims
a brochure currently being distributed by the Department
of Energy. "The deuterium in the world's oceans, if allowed to undergo fusion little by little, would supply man-
kind with enough energy to keep going at the present rate
for billions of years."

It is easy to believe the dreamers promoting technological pacifiers to a public battered by our energy crisis. Yet
if anything should be learned from the present nuclear
debacle, it is that we need to be careful about our fantasies.
Its time for more realistic dreams, based on the realities
facing us.

Through conservation, appropriate technologies, and
new investment in efficient technologies, a workable future is possible. It would not feature artificial suns at baseball parks, but it would permit us to lead a comfortable
existence within our means. Maybe that is all we should
expect.

Three Mile Island: The Loss of Innocence

Ellyn R. Weiss

Three Mile Island lies in the middle of the Susquehanna River in the rural countryside of central Pennsylvania. Several small towns encircle the island, including Middletown, Royalton, Goldsboro, and Falmouth. Hershey, prosperous home of the chocolate factory, is about 10 miles from the island. Despite the rural character of the area, it borders on a major population concentration. The metropolitan Harrisburg area, including the state capital and over 600,000 residents, is only some 12 miles up the river.

The island is the site of two nuclear power reactors owned and operated by the Metropolitan Edison Company, known locally as Met Ed, a subsidiary of General Public Utilities (GPU), a larger company with interstate holdings. The pressurized-water reactors were designed and manufactured for Met Ed by the Babcock and Wilcox Company. The second of these reactors began full commercial operation on December 31, 1978, just barely (and

Goldsboro, Pa. Looking east across Goldsboro Square at the Three Mile Island plant. Photo by Dennis Gordon, Akron Beacon Journal.

not coincidentally) in time to qualify Met Ed for a multimillion dollar tax writeoff for the year. On March 28, 1979, less than three months after beginning commercial operation, Three Mile Island Unit Two had the most serious nuclear accident ever to happen in the 25-year history of the civilian nuclear power program.

The plant's enormous cooling towers, which dominate the local landscape, were adopted as a visual symbol of the accident by the television crews from around the world that quickly gathered to monitor the minutest details of the plant's progress toward either recovery or meltdown. Almost overnight, Three Mile Island's (TMI) cooling towers became the international emblem for nuclear power, undoubtedly causing great unhappiness among the nuclear industy. The new symbol, looming and ominous, graphically reflects a loss of innocence about nuclear power that is fully justified by the events at TMI.

THE ACCIDENT

Things seemed to be under control as the shift operators at TMI prepared for sunrise and a day's sleep. Outside, the pre-dawn quiet was broken only by occasional cars and trucks and the moos of cattle being let out to pasture by early-rising farmers. The silence was shattered by a roar of steam at about 4 A.M. A few people living near the plant were awakened by the noise. Inside the plant, a double-tone alarm roused the crew. Within two minutes, 200 alarms sounded in the control room. "I'd like to have thrown away that alarm panel," control room operator Edward Frederick would say later. "It wasn't giving us any useful information." While confusion reigned in the control room and plant personnel converged there, an un-precedented series of events proceeded, step by step, toward the worst accident in the history of U.S. commercial nuclear power plants.

Nuclear Regulatory Commission (NRC) sources have theorized that it seems to have started with a little bit of water—out of the tons of water circulating through the plant's massive pipes—going into an air tube in the plant's secondary cooling system, also called the feedwater system. That cooling system circulates water to a steam generator, which draws heat from the primary cooling system, the one that actually cools the more than 100 tons of uranium fuel rods in the reactor's core. The unwanted water crept into a tube connected with feedwater pumps. The moisture in the air system caused a safety valve to close, which in turn caused automatic shutdown of the feedwater pumps.

Up to this point, what had happened could hardly be called an accident. Feedwater failures, while undesirable, are common. And the next two automatic plant responses were also common and appropriate.

Control room of nuclear power plant. Photo by Daniel S. Brody.

First, the turbines tripped, or shut down. The turbines are the part of the plant that generate electricity. Secondary coolant is boiled into steam as it passes through the steam generator. That steam spins the turbines, where magnets rotating inside coils of wire create electric current, which is fed to customers via power lines. Turbine trips are fairly routine occurrences at nuclear power plants.

The second automatic response to the failure of the feedwater pumps was the startup of three backup emergency feedwater pumps. If everything had gone normally, within seconds feedwater would have been recirculating back to the steam generators to complete another circuit of the secondary cooling system. But there were two problems. Someone had closed the two valves through which the emergency feedwater was supposed to flow. This incredible breach of procedure, a flagrant violation of the NRC's safety rules, prevented the feedwater from doing its job in cooling the reactor.

Plant maintenance workers would later tell the presidential commission investigating the accident that they had left the valves open after tests a few days before the accident. They testified that the records showing that the valves had been properly left open were thrown out, as they always were. No matter. Without secondary coolant to draw away some of the primary coolant's heat, temperatures and pressures in the reactor began to climb dangerously.

Within six seconds, two special relief valves in the primary coolant system opened to ease the pressure. The "electromagnetic relief valves" sit atop the reactor's pressurizer—an electrically heated tank of water that plant operators use to regulate pressure in the primary coolant system. That coolant loop is kept under high pressure to prevent the superheated water from boiling. Highly pressurized primary coolant courses through the core, drawing heat away from the fuel rods and, through the steam generator, passing some of that heat to the secondary coolant system, which drives the power-generating turbines. Primary coolant then recirculates back to the core to repeat the process.

With the loss of secondary coolant forcing pressure up in the primary loop, the opening of automatic pressure relief valves was the way to bring pressure under control. However, the pressure continued to rise for a number of seconds, activating an automatic reactor shutdown or SCRAM. Eventually, the pressure began to come back down to normal levels. Unfortunately, one of the relief valves stuck open after the pressure had dropped, draining vital coolant from the fuel rods. The failure of the relief valve to shut seriously aggravated the accident.

That stuck valve was like a hole in the primary coolant system, having the same effect as a small pipe break and leading to what nuclear engineers call a Loss-of-Coolant

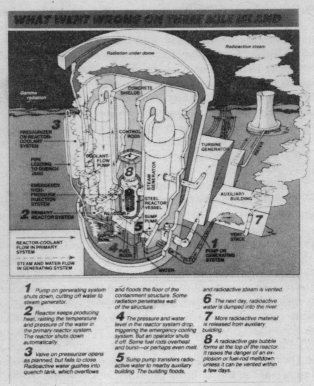

WHAT WENT WRONG ON THREE MILE ISLAND

1 Pump on generating system shuts down, cutting off water to steam generator.

2 Reactor keeps producing heat, raising the temperature and pressure of the water in the primary reactor system. The reactor shuts down automatically.

3 Valve on pressurizer opens as planned, but fails to close. Radioactive water gushes into quench tank, which overflows

and floods the floor of the containment structure. Some radiation penetrates wall of the structure.

4 The pressure and water level in the reactor system drop, triggering the emergency cooling system. But an operator shuts it off. Some fuel rods overheat and burst—or perhaps even melt.

5 Sump pump transfers radioactive water to nearby auxiliary building. The building floods,

and radioactive steam is vented.

6 The next day, radioactive water is dumped into the river.

7 More radioactive material is released from auxiliary building.

8 A radioactive gas bubble forms at the top of the reactor. It raises the danger of an explosion or fuel-rod meltdown unless it can be vented within a few days.

Accident (LOCA). Steam and water poured out of the valve, and reactor pressure dropped until it triggered yet another automatic safety device: the Emergency Core Cooling System (ECCS) started up and began to force cold water into the reactor. But the hole in the system remained, eventually allowing hundred of thousands of gallons of highly radioactive water to spill onto the floor of the high-domed concrete containment building that houses the reactor. Sump pumps, operating automatically during the chaos of the event, pumped the water out of the radiation-shielded containment into an auxiliary building, where a faulty tank periodically "burped" radioactive elements into the air. This was one of the major sources of radioactivity released to the public.

While the sump pump churned away, operators in the control room turned the ECCS off. In the weeks following the accident, this would be pointed to as a key "operator error" that contributed to making the mishap much more serious than it otherwise would have been. Actually, the operators seem to have made the responses they were trained to make, although instruments were giving them a false picture of what was happening inside the reactor. They were relying on a meter that reads the water level in the pressurizer. Those readings indicated the presence of sufficient water in the reactor to keep the core covered. Operators are trained to assume that pressurizer levels adequately reflect the level of coolant in the reactor. Assured that those levels were sufficient, they switched off the ECCS. The people in the control room had no way to directly gauge the water level where it counted most—in the reactor core. Such gauges are not even a federal licensing requirement.

The primary goal of nearly all safety systems in a reactor is to keep the reactor cooled. If deprived of enough cooling water, the fuel rods inside the core overheat and start to melt. As the sequence progresses, the fuel rods sag together, and the core melts through the containment and into the earth, releasing enormous amounts of radioactivity into the environment.

As the chronology of the early part of the accident demonstrates, all those involved were moving in uncharted waters as they attempted to cope with an accident that was not anticipated by the federal regulatory process supposedly safeguarding the public. The sequence was not predicted by any of the industrial engineers. Nor was its deadly potential identified by the NRC's own engineers when the plant received its license to operate. As Dr. Roger Mattson, assistant director of nuclear reactor regulation for the NRC, put it, "We saw failure modes, the likes of which have never been analyzed."

Eight minutes after the feedwater blockage created all these problems, plant operators had discovered the closed valves and opened them, releasing a rush of cold water into the steam generators, which by this time had virtually boiled dry. It was too much for the steam generators to take. Leaks developed between the primary and secondary systems, releasing radioactivity into the secondary system.

One of the purported advantages of the pressurized-water reactor design is the existence of two cooling systems to prevent the radioactive primary coolant from leaking to the atmosphere. With the steam generator barrier thus broken down, radiation soon escaped into the environment. Eventually one steam generator was closed off to plug the leaks of radioactive water.

With the ECCS shut off and the pressurizer valve continuing to spill radioactive primary coolant onto the containment floor, pressure levels in the reactor continued to drop. About 11 minutes into the accident, plant operators restarted the ECCS. Things continued to go wrong. Soon the massive pumps responsible for circulating primary coolant through the core began to vibrate dangerously. Despite the best efforts of the ECCS, the open pressurizer valve created gaps in the coolant flow. These "voids" were made up largely of steam, something the coolant pumps are not designed to handle. As they wrestled with the vapor, they shook dangerously, threatening to break apart or separate from the pipes, potentially causing a full Loss-of-Coolant Accident. With that fear in mind, the people running the reactor turned the big pumps off more than an hour into the mishap. This act too was cited as an "operator error" by officials who sought to absolve themselves and their technology of blame. Had there not been a hole in the system, the operator's action might have helped things. As it was, shutting off the main pumps may have been the worst thing they could have done.

Within minutes, core temperatures went off the scale at 750 degrees Fahrenheit—150 degrees above the core's normal operating temperature. For hours the plant's computer printed question marks instead of temperatures. A technician taking readings from in-core thermometers said he was getting readings of 2300 degrees. He told his superiors he did not believe what he was seeing, and his superiors apparently took him at his word.

The core became uncovered—partially exposed without coolant—and the fuel rods began to sag and rupture. The full extent of the core's damage will not be known until it can be inspected. With several feet of intensely radioactive water on the floor of the containment building and lethal levels of radioactivity surrounding the reactor vessel, it will be months or years before core damage is assessed. A major factor in mitigating the accident was the lucky fact that the core was relatively young, with a minimal buildup of hot nuclear-fission products.

At about 6:20 A.M., technicians got around to closing the hole in the primary coolant system by using a bypass valve to "isolate" the stuck pressurizer relief valve. Had operators in the control room taken their instruments a little more seriously, they might have acted sooner. A report by TMI's owner, General Public Utilities, later revealed that a gauge showed temperatures at the stuck valve running 155 degrees above the safety shutoff level. Operators said the valve normally operated above normal temperatures, so they discounted the importance of the readings. TMI

Unit Two superintendent Joseph Logan told the presidential commission, "I had been living with a leaky relief valve for quite some time." So had the people downwind of the reactor.

Pressures and temperatures in the core continued to fluctuate. Parts of the core were uncovered for about 15 hours. Temperatures in the core soared well past the safe operating level. When the metal tubes that hold the uranium fuel pellets in the core sagged and split, deadly fission products were released into the primary coolant, the containment building, and, ultimately, the atmosphere. As the rods drooped together, their bunching further hampered the flow of coolant. A meter at the top of the containment dome registered radiation levels far in excess of the lethal radiation dose for human beings.

Tony Auth. © 1979, The Philadelphia Inquirer. The Washington Post Writers Group.

THE COMING OF THE BUBBLE

Disbelief was a common emotion during the early days of the accident. At first, it was disbelief that there had actually been an accident. Then there was disbelief that it was serious. Disbelief was also felt by local and state officials and the hordes of journalists converging on the site. As they found themselves being fed conflicting information by the plant's operators and by NRC officials, their disbelief became hostile skepticism.

Early on Friday, the third day of the accident, official radiation monitors began to pick up sudden releases—the "uncontrolled emissions" that instilled the first full breath of fear into the event. Met Ed and federal officials publicly quibbled over who was boss at Three Mile Island. The utility at one point threatened to dump the crisis into Washington's lap. The public was completely confused by the time the real headache—an elusive gas bubble—emerged on the scene.

"The Bubble" would occupy the nation's, and the world's, attention through the weekend. It was obstinate, ever changing, of unknown composition, and it apparently wasn't going anywhere. Like so many other aspects of the Three Mile Island accident, the real story of the bubble will probably never be known.

The public first learned of the bubble on Friday. Making complex extrapolations from temperature and pressure readings of the reactor, and chemical analysis of some of the spilled coolant water, technicians at the site determined that there was a hydrogen bubble poised over the fuel rods in the core of the reactor.

Officials thought the bubble may have been generated by a chemical reaction between the melting zirconium surrounding the fuel rods and the primary coolant. Estimates of its size and precise composition shifted almost from hour to hour. The danger posed by the bubble was all too clear. If it grew, as it at first seemed to do, it could push coolant out of the core, reexposing the rods and reviving the possibility of a meltdown. If the hydrogen exploded or burned, it could shatter the reactor vessel and perhaps rupture the huge 4-foot-thick steel-and-concrete containment building—the final barrier between the melting core and the environment.

Most people read the newspaper accounts, watched the television specials on the escalating crises, and were forced to consider a peaceful atomic disaster. Hundreds of nuclear engineers and computer specialists huddled night and day, developing models of the disabled reactor and trying to come up with ways to get rid of the bubble. Soon scenarios for dissolving the bubble were filtering through the press and down to a baffled public.

Just when it seemed someone was going to have to make some serious choices about what to do, the bubble vanished. On Monday, it was said to have shrunk. By Tuesday it was said to have disappeared. And as Met Ed officials tried to claim credit for the favorable turn of events ("Based on the game plan, this is exactly what we expected to happen," beamed a utility engineer), government laboratories began working around the clock to figure out how the bubble went away. One top NRC aide said it may have leaked out through some cracked pipes. But most officials sought to give credit to a jury-rigged technique of manipulating reactor-vessel pressure to force the hydrogen out of the reactor into the containment and then rendering it harmless with mechanical "recombiners."

Finally there is an ongoing effort to have the troublesome bubble erased from the official record. NRC officials have testified to a congressional panel that a miscalculation led to a "mistake"—a belief that a bubble existed when in fact it did not. It is at least possible that the bubble, which

Tony Auth. © 1979, The Philadelphia Inquirer. The Washington Post Writers Group.

so riveted official and public attention, was a phantom. We may never know.

THE WARNING SIGNS

Even though Three Mile Island Unit Two was among the newest reactors in the nation when the accident occurred, the plant had a history of serious technical problems, many of which popped up simultaneously early that March morning. In one of the little ironies of the mishap, the reactor's core went critical—sustained its first nuclear chain reaction—on March 28, 1978, one year to the day before TMI became a household word. The day after it was first operated, a fuse blew, forcing the reactor to shut off automatically. One of the relief valves on top of the pressurizer stuck open, and the ECCS came on to keep the core covered—just as it did a year later.

Technical problems plagued the new reactor all through 1978. It had to be shut down on October 13 for repair of a broken pressurizer valve. On November 3 a plant operator incorrectly left feedwater valves closed, causing a loss of secondary coolant and a reactor shutdown. Another reactor shutdown caused by loss of feedwater took place four days later, this one punctuated by an ECCS startup. All told, the plant was shut for repairs 195 of the 274 days between its startup and the day it entered service.

Feedwater problems continually hampered efforts to bring the plant into commercial use by year's end—something General Public Utilities desperately wanted in order to qualify for enormous tax and rate advantages available under federal and state law.

A study by Ralph Nader's Public Citizen organization lists the powerful financial incentives that prompted GPU to force the plant into commercial operation despite repeated safety problems, with the tacit approval of the Nuclear Regulatory Commission. GPU stood to gain between $25 and $40 million in federal investment tax credits by bringing the plant into use before tax year 1978 ran out. TMI Unit Two's owners also stood to gain about $20 million in depreciation writeoffs by putting the plant in service. And once the reactor was generating electricity, its capital cost could be included in the utility's rate base. During the first three months of 1979, the GPU subsidiaries that drew power from TMI were granted rate increases of almost $100 million. All this money was for a power plant that had proved only that it was unreliable and unsafe.

NRC had granted the plant an operating license on February 8, 1978, with assurances from Met Ed, the utility that held the license, that the plant would be thoroughly tested before it was put in service. "When the NRC grants an operating permit," NRC Operating Data Section Chief Richard Muranaka told Public Citizen, "the company is authorized to operate. When the company puts the plant into commercial operation really depends on their tax structure. There's some tax advantage for a plant to go into operation before the end of the year. This is particularly evident when a plant goes into operation late in the year."

Met Ed declared TMI Unit Two to be in commercial operation at 11 P.M. on December 30, 1978—only 25 hours before the tax year ended.

PORTRAIT OF A FAILURE

Perhaps the most explicit recounting of the chaotic climate in which decisions were being made is provided by the public records of the NRC's deliberations as the crises escalated. The five commissioners who run the agency—Chairman Joseph Hendrie, Victor Gilinsky, Richard Kennedy, Peter Bradford, and John Ahearne—did not even hold an official meeting on TMI until three days into the accident. Up to that point, NRC's response was limited largely to the dispatch of inspectors to the plant site a few hours after first word of the mishap reached Washington.

A federal law known as the "Government as the Sunshine Act" requires the NRC and other agencies to keep a record, including minutes or a transcript, of all "meetings," which is defined loosely by the NRC as all occasions on which three or more of the NRC commissioners get together to discuss official business.

On Friday, March 30, the five NRC commissioners began a series of meetings closed to the public that continued on a more or less constant basis for almost a week. NRC employees with tape recorders literally trailed after the commissioners, turning the recorders on whenever three of them converged in one spot. These were transcribed and eventually made public in response to numerous outside requests filed immediately under the Freedom of Information Act.

Transcripts of the commissioners' meetings depict, on the third day of the accident, an agency in utter confusion, with only rudimentary and unreliable communication with Met Ed, Pennsylvania state authorities, and its own people at the site. The immediate question facing the commission at the beginning of the transcripts was whether to recommend to Pennsylvania Governor Richard Thornburgh that he order an evacuation of the more than one million people within 20 miles of TMI. An uncontrolled release of radiation Friday morning that was much larger than previous emissions heightened the alarm. The following are quotes from the official transcripts:

MR. JOSEPH FOUCHARD (NRC information officer): This is Joe, Mr. Chairman, I just had a call from my guy in the governor's office and he says the governor says the information he is getting from the plant is ambiguous, that he needs some recommendations from the NRC.

MR. HAROLD DENTON (director of reactor regulation for NRC): It is really difficult to get the data. We seem to get it after the fact. They opened the valves this morning, or the let-down, and were releasing at a six curie per second rate before anyone knew about it. By the time we got fully up to speed, apparently they had stopped, there was a possible release on the order of an hour or an hour and a half. . . .

We calculate doses of 170 millirems per hour at one mile, about half that at two miles and at five miles about 17. Apparently, it is stopped now, though I'd say there is a puff release cloud going in the northeast direction and we'll just have to see. We did advise the state police to evacuate out to five miles but whether that has really gotten pulled off, we'll just have to—

MR. FOUCHARD: Well, the governor has to authorize that, and he is waiting for a recommendation from us. . . .

CHAIRMAN HENDRIE: Harold, where is— For a puff release, what you have got is an oblong plume headed out. Where is it now, would you guess, that is, if we go

ahead and suggest to the governor that the evacuation
in that direction out to five miles be carried out, is it
going to be after the fact of the passage of the cloud?

MR. DENTON: Well, if they haven't gotten it cranked
up, it might well be after passage. There are people
living fairly close to the northeast direction. I guess the
plume has already passed there. . . .

Yes, I think the important thing for evacuation to get
ahead of the plume is to get a start rather than sitting
here waiting to die. Even if we can't minimize the individual dose, there might still be a chance to limit the
population dose. . . .

It just seems like we are always second, third hand;
second-guessing them. We almost ought to consider the
Chairman talking to the owner of the shop up there and
get somebody from the company who is going to inform
us about these things in advance if he can and then what
he is doing about it if he can't. We seem not to have
that contact.

COMMISSIONER GILINSKY: Well, it seems to me we
better think about getting better data. . . .

MR. FOUCHARD: Don't you think as a precautionary
measure there should be some evacuation?

CHAIRMAN HENDRIE: Probably, but I must say, it is
operating totally in the blind and I don't have any confidence at all that if we order an evacuation of people
from a place where they have already gotten a piece of
the dose they are going to get into an area where they
will have had .0 of what they were going to get and now
they move someplace else and get 1.0.

COMMISSIONER GILINSKY: Does it make sense that
they have to continue recurrent releases at this
time . . . ?

MR. DENTON: I don't have any basis for believing that
it might not happen—is not likely to happen again.

I don't understand the reason for this one yet. . . .

" THE SITUATION IS WELL IN HAND! "

CHAIRMAN HENDRIE: Now, Joe, it seems to me I have got to call the governor—

MR. FOUCHARD: I do. I think you have got to talk to him immediately.

CHAIRMAN HENDRIE: —to do it immediately. We are operating almost totally in the blind, his information is ambiguous, mine is nonexistent and— I don't know, it's like a couple of blind men staggering around making decisions.

Chairman Hendrie then had a call placed to Governor Thornburgh. Since the radioactive cloud—which Hendrie described as containing "a pretty husky dose rate"—had already passed over the population, and NRC was "behind the event," Hendrie did not recommend an evacuation. Instead, he suggested that the governor advise people within 5 miles to stay indoors. Governor Thornburgh pressed Hendrie to clarify this recommendation in light of the fact that NRC personnel at the site had advised an evacuation earlier that morning. Thornburgh asked if the evacuation recommendation was justified. Hendrie said he didn't know.

The governor then advised his constituents living around TMI to stay indoors at about 10:30 A.M. Less than two hours later, in a nationally televised press conference, Thornburgh took the more serious step of advising children of preschool age and pregnant women living within 5 miles of the plant to evacuate. More radiation had been released. Perhaps more significantly, by this time the first public mention had been made of the potentially explosive hydrogen bubble that the engineers believed to have formed at the top of the core. The bubble would preoccupy the engineers for the next five days. Any precipitate action or some unforeseen event might cause it to burn or explode. It was all too frighteningly easy to construct a scenario culminating in meltdown. Under these circumstances, the NRC's Roger Mattson began to recommend precautionary evacuation in the early afternoon of Friday.

> MR. MATTSON: The latest burst didn't hurt many people. I'm not sure why you are not moving people. Got to say it. I have been saying it down here. I don't know what we are protecting at this point. I think we ought to be moving people.

If the transcripts show the NRC unable to comprehend the accident or manage the crisis, they are hardly more reassuring on the competence of the organization actually in control of the reactor, the Metropolitan Edison Company. After Denton arrived on the scene, he began to relay back to Washington some very disturbing observations on the subject:

> MR. DENTON: The utility is a little shy, in my view, of technical talent. We outnumber them. They are pretty thin. I'm trying to convince them to bring in compara-

ble levels from their own organization. Their coopera-
tion is good, but it is obvious that they are a small outfit
here and the guys are getting swamped with demand.

On Saturday, April 31, Denton's concern is still present,
though stated in his peculiarly soothing style:

> MR. DENTON: I guess I've developed a management
> concern about the capability of the utility here to cope
> with new problems that come up. They've stretched
> very thin in some areas . . . [I] think they need stem-to-
> stern reinforcements down here in many areas. . . .
>
> I would sure like to see them muster their resources
> . . . and tackle these problems clearly. . . .
>
> If you ask them what happens if, you know, the atti-
> tude is well, maybe that won't happen and if it does,
> we'll cope with it then. . . .
>
> It's just too low a level of attention.

Tony Auth. © 1979, The Philadelphia Inquirer. The Washingtyon Post Writers Group.

Technology of Nuclear Power

Technology of Nuclear Safety

Chairman Hendrie agreed to place a call to the president of Met Ed, presumably to persuade him to raise the "level of attention" being given to TMI.

Meanwhile, Met Ed itself had been presenting a radically different picture to the public. Its posture from the beginning had been to minimize the risk from the unstable reactor and project a confident, "in-charge" image. This contrasted with Chairman Hendrie's assessment of the utility as "not all that strong technically."

On Wednesday at 3:30 P.M., during the first day of the crisis, Met Ed Vice-President John Herbein termed the accident "nothing we can't take care of." He also said that day that the hazard to the public was "minimal" and that the plant's systems had "functioned the way they were supposed to." On Thursday, Herbein claimed that the plant was "in stable condition and operating normally." And on Friday, as events escalated dramatically, Herbein said, "The situation is under control and yes, we know what we're doing. And shortly the plant will be in a more stable condition than it is now."

Met Ed was later taken to task severely by the press, who sensed an attempt by the utility to cover up in the contrast between Met Ed's bland reassurances and the NRC's substantially more sober assessments.

Public officials in several towns near the plant site were outraged that Metropolitan Edison officials had not notified them of radiation releases that threatened their communities. Most were never contacted. The mayor of nearby Middletown, Robert Reid, did talk to company officials, who said there was "no problem." He was not told about the radiation leaks.

The Nuclear Regulatory Commission did not become deeply involved in the management of the crippled plant until two days into the accident. After the large, unplanned (according to the NRC) release of radioactivity from the plant on Friday, March 30, President Carter ordered the agency to take over both decision-making and public-information tasks concerning the attempts to bring the reactor under control.

This announcement was made on Saturday, when the company and the NRC held back-to-back press conferences. It was clear that there was tension between the two groups. Met Ed Vice-President John Herbein said the troublesome hydrogen bubble had shrunk. But Harold Denton, the NRC's top official at the scene of the accident, then directly contradicted Herbein.

President Carter took a sudden personal interest in the safety of nuclear power plants. He arranged to speak daily with top NRC officials at the site. In an attempt to restore public confidence in the midst of what appeared to be a chaotic federal response, Carter also made a special trip to inspect the reactor site on Sunday, April 1. He spent 36 minutes in the plant and then announced to television cameras that he would be "personally responsible for thoroughly informing American people about this incident."

Carter further attempted to bolster public confidence by establishing an independent presidential commission to investigate the causes of the accident. Named to head the commission was John Kemeny, president of Dartmouth College in Hanover, New Hampshire, and former staff mathematician on the Manhattan Project, which developed the first atomic bomb. The commission is in the process of holding hearings as this book goes to press.

Pennsylvania Governor Richard Thornburgh, who was installed in office only three months before the accident, struggled with a political and moral dilemma. If he failed to order residents near the plant to evacuate and a catastrophic accident occurred, his political career could be finished, to say nothing of the lives of the residents. On the other hand, such an unprecedented evacuation would pose serious risks of its own.

Marlette. Copyright The Charlotte Observer. King Features Syndicate.

Overall, the performance of the officials charged with managing the nuclear emergency revealed three things: no one in authority knew what the situation was, no one *could* know what the situation was because of inadequate information, and, still worse, no one knew what to do to protect the public. This led to a confusing range of statements on the severity of the accident and widespread public skepticism that the truth was being revealed. A *New York Times* editorial on March 30, headlined "The Credibility Meltdown," asked, "Was it a little leak, a bigger leak, or a general emergency? The reactor's operators said one thing, state officials another, federal officials yet another, not to mention the contribution of equipment manufacturers and politicians."

THE AFTERMATH

Before Three Mile Island, the nuclear industry claimed
that it had succeeded in taming the nuclear genie; the risks
of a serious accident threatening public health and safety
were negligible. The regulators agreed. As one measure of
this faith, NRC has never required that the states in which
nuclear plants operate have evacuation plans. As another,
the NRC simply deemed a major reactor accident such as
a meltdown to be so unlikely as to be essentially incredible.
Therefore, the NRC refused to consider the potential
catastrophic consequences of such accidents in the Environmental Impact Statement required for each plant. As a
third measure of faith, the NRC has allowed plants to continue operating in the face of known safety deficiencies
such as vulnerability to the effect of fires.

Congress appeared to be largely indifferent. It had risen
up in 1974 and broken the old Atomic Energy Commission
(AEC) into two parts, separating its overtly promotional
functions from its regulatory functions. The NRC was supposed to be regulator only. But the NRC staff was reincar-
nated from the AEC, and the NRC's first official act was to
adopt all of the AEC's rules and regulations. Congress
seemed content.

On April 27, 1979, the NRC ordered all but two operating Babcock and Wilcox nuclear plants to shut down immediately, pending modifications to the plants and retrain-
ing of operators.

It is clear that major changes will be wrought in the
months ahead both in the design of plants and the training
of operators and also in the nuclear program and the relationship between licensees and regulators. Chairman Hendrie told Congress on April 10, 1979 that "[w]e cannot
have an acceptable nuclear power program in this country
if there is any appreciable risk of events of the TMI kind

occurring." It is possible that changes required to remove that risk will tip the economic balance irrevocably against nuclear power.

Whatever may be the final fate of nuclear power, the people who live near Three Mile Island will bear the poisoned legacy of the accident. Long after the mechanical and human failures at TMI are fully understood, the unseen effect of the radiation will be a matter of great debate.

The amount of radiation released will never be precisely known. Following the accident, Secretary of Health, Education, and Welfare Joseph A. Califano, Jr., testified before Congress, "The facts on the extent of exposure remain uncertain." He said this was the case because,

> . . . during the first three days after the accident, when the releases were the highest, fewer than 20 ground level dosimeters were in place; after that time, there were additional NRC and Food and Drug Administration dosimeters. But some areas had no dosimeters and exposure calculations had to rely entirely on extrapolation; moreover, it is uncertain how many persons were located in each area.

Official discussion of the radiation perils posed by leaks at TMI limited itself to "puff releases" of "inert gases" like xenon and krypton, and cursory discussion of iodine-131. There were also constant comparisons to familiar dental x-rays. Those laymen who plowed through the morass of conflicting numbers and esoteric language were encouraged to believe that the radiation emissions—conceded to be the worst from any commercial nuclear power plant— did not pose a significant threat to public health. Secretary Califano took this position when he appeared before Congress and, via TV news crews, the nation, saying that his advisers believe that no more than ten cancers would develop within a 50-mile radius of the plant.

When Califano reported to Congress a month earlier, on April 4, the estimate of total population dose, derived from NRC figures, was 1800 person rems. As of mid-April, that had doubled to approximately 3500 person rems. The secretary added on May 3 that he expected the revised estimates of total dose "to increase again."

CONCLUSION

March 28, 1979, was a watershed for nuclear power. Before then, it was possible to believe, as many people did, that an infallible technology had been fashioned to bring us the benefits of the "peaceful atom." The sanguine claims of infallibility have certainly been shattered.

The nuclear industry would like to replace the dead myth of infallibility with a new one, the myth of necessity. According to this, the American way of life depends on nuclear energy. These self-serving claims do not stand up to scrutiny (see "The Heart of the Issue: Money" for details). The growth in electricity demand has decreased sharply since 1972, causing a virtual standstill in orders for new plants. Even if nuclear power were to grow at the vastly inflated rates predicted during the Nixon-Ford era, it simply does not have the potential to be more than a minor energy source. This is fortunate. It makes disengagement from a dangerous nuclear future a relatively painless prospect.

Splitting Atoms
to Boil Water

———

George R. Zachar

Jargon has many uses. For those who know it, jargon is a convenient shorthand. For those who don't know it, jargon can be an impenetrable wall of buzz words that effectively hide meaning. Before the accident at Three Mile Island, the jargon of nuclear power had an almost mystical aura. Feedwater loops, steam generators, and pressurizers were left to the nuclear engineers. Most people used electricity, paid their escalating utility bills, and griped about the cost. The battle against nuclear energy—a battle that had been going on for years in courtrooms, at regulatory proceedings, and at power-plant construction sites—was a sideshow occasionally reported in the press. Now nuclear power in all its complexity is at the center of a national debate. Three Mile Island put nuclear power on the political map.

This is an atlas.

The nuclear power controversy boils down to two questions: Is it safe? Is it economical? It is neither. This chapter is a guide to the life-menacing technical side of atomic power.

The accident at Three Mile Island posed an immediate

threat to hundreds of thousands of people. Soothing assurances by the plant's owner that all was well did nothing to
prevent more than 50,000 people from fleeing the region.
They had the right idea. No one—not the people running
the plant, not the government, not the hundreds of "experts" airlifted to the scene—knew what was going on
inside the reactor core, where a hundred-odd tons of uranium fuel threatened to go out of control and melt its way
into the Pennsylvania soil.

For almost a week, news of the impending nuclear disaster near Harrisburg dominated the media. Try as they
might, the plant owner (Metropolitan Edison) and a
quickly assembled cavalry of nuclear supporters could not
defuse public skepticism. The nuclear-power advocates
have been forced to fall back on their flimsy economic
justifications. A look at the nuts and bolts of nuclear power
shows why it can never be safe.

SIGN LEFT ON A PORCH BY TMI: "GONE FISSION"

Nuclear power plants differ from coal- or oil-fired power
plants in only one major respect. They use the **fissioning,**
or splitting, of atoms as a heat source instead of the burning
of a fossil fuel. In most electricity-generating plants, heat
is used to boil water into steam which is used to spin a
turbine. The turbine rotates a magnet in a coil of wires,
generating the electricity, which is fed by power lines into
homes, factories, and offices. Nuclear power—with all its
long-lived radioactive hazards—is simply one way to boil
water.

Everything we touch is made of **atoms**—the invisible but
intricate "building blocks of matter." Albert Einstein in
1905 mathematically determined that matter—that is,
atoms—could be converted into energy. His landmark
equation $e = mc^2$ expresses that relationship. Energy
released *(e)* is equal to a given quantity of mass *(m)* multi-

The Atom

Atomic power gets its name from the atom—the basic component of matter. Virtually everything in the universe is made up of atoms. And atoms are themselves made up of a heavy, central **nucleus** surrounded by a cloud of orbiting **electrons**. The nucleus, which accounts for most of an atom's weight, is made up of two types of particles: positively charged **protons** and non-charged **neutrons**. The electrons form a kind of shell around the nucleus, bearing negative charges to counter-balance the positive charges of the protons. Each electron weighs about 1/1800 what a neutron or proton weighs, so an atom's **atomic weight** is calculated by tallying the sum of its protons and neutrons. Different elements are distinguished by the number of protons—called the **atomic number**—in their nuclei. Most of an atom is space. To put it in perspective, if a nucleus was an inch across, the orbit of its electrons would have a radius of four miles.

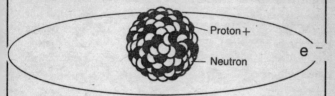

Proton +

Neutron

e −

With all these different types of particles and charges coming together on an atomic scale, there are bound to be problems. The nucleus, for example, has a major structural difficulty. Because like charges repel one another, a delicate structural balance between protons and neutrons must be maintained or the nucleus, and the atom, will fly apart. As atoms get bigger, more and more positive charges are jammed into the nucleus, and the atom becomes increasingly unstable. More nuclear **binding energy** is needed to hold it together. Tension between the positively charged nucleus and its negative charge-bearing electron cloud is another potential source of atomic instability. Unless the charges equal each other and cancel themselves out, the atom will be unstable. As if all this were not enough, the number of neutrons in a nucleus can vary. Atoms with identical numbers of protons (atomic numbers) but different numbers of neutrons are called **isotopes**. Uranium has many isotopes—all have 92 protons, but one, with an atomic weight of 235, has 143 neutrons. Another has 146 neutrons for an atomic weight of 238. These isotopes are called U-235 and U-238, respectively.

Radiation

The bigger the atom, the more likely it will be unstable. Many large atoms, particularly those bigger than lead (which has an atomic number of 82) can move toward greater stability by emitting **radiation**—casting off little pieces of themselves to bring their various charges and weights into balance. This process is called **radioactive decay**.

Different radioactive atoms—called **radioactive isotopes** or **radionuclides**—emit different types and quantities of radiation as they move toward stability. Many atoms emit **ionizing radiation**—radiation that can alter the atomic structure of any atoms it comes in contact with. This is how radiation can hurt or kill living tissue. By disrupting the atoms in living cells, in gene-bearing chromosomes, for example, ionizing radiation can impair life-maintenance activity, cause genetic defects, trigger wild cellular reproduction (called **cancer**) and in many cases, kill cells outright. There are three types of ionizing radiation: **alpha particles, beta particles** and **gamma rays**. Alpha particles are cast off by a radionuclide's nucleus. They consist of two protons and two neutrons, just like a complete helium atom nucleus. This relatively large particle cannot penetrate the skin, but can cause skin cancer. Beta particles, which are simply electrons, can penetrate your body an inch or more. Gamma rays, which are a higher-energy version of X-rays, can pass right through you, shattering atoms in living cells along the way. And, the ingestion (inhalation or swallowing) of alpha- or beta-emitting substances can be extremely dangerous. **Alpha-** or **beta-emitters** can lodge in the body, and bombard nearby cells with relatively enormous doses of ionizing radiation, with the potential for triggering cancer months or years after ingestion. For example, inhalation of a millionth of a gram of **plutonium-239**, an intense alpha-emitter produced in ordinary nuclear reactors and used in nuclear bombs, can induce lung cancer. One pound of Pu-239, evenly distributed, could kill every human on Earth.

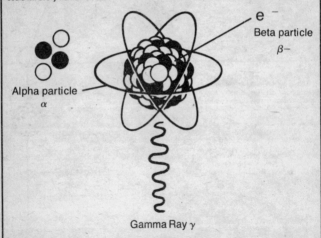

e⁻
Beta particle
$\beta-$

Alpha particle
α

Gamma Ray γ

Different radionuclides decay at different rates. The rate of radioactive decay is measured in **half-lives**—the amount of time it takes half a quantity of a radioactive isotope to decay into other atoms, called **decay products**. Sometimes, the products are themselves radioactive, with their own half-lives. One pound of strontium-90, a radioactive poison with a half-life of roughly 30 years, will decay into a half-pound of strontium and a half-pound of decay products in 30 years. In another 30 years, there will be a quarter-pound of strontium. In another three decades, it will be an eighth of a pound, and so on. The standard yardstick for gauging how long a radioactive isotope remains dangerous is to multiply its half-life 10 to 20 times. After 10 half-lives, one-thousandth of the original poison would remain. After 20, it would be one-millionth. Plutonium-239, one of the most deadly by-products of nuclear power plants, has a half-life in the neighborhood of 24,000 years, so it is lethal for a quarter-million years or more.

Fission at a Glance

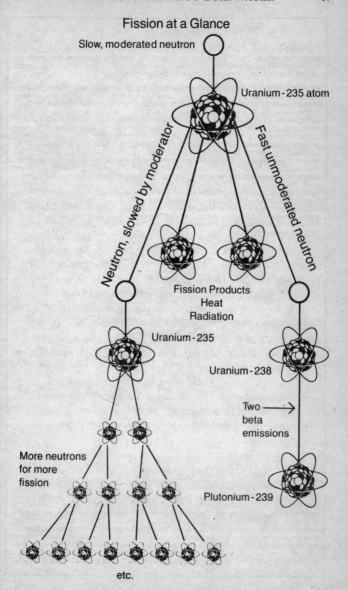

Slow, moderated neutron

Uranium - 235 atom

Neutron, slowed by moderator

Fast unmoderated neutron

Fission Products
Heat
Radiation

Uranium - 235

Uranium - 238

Two →
beta
emissions

More neutrons
for more
fission

Plutonium - 239

etc.

plied by the speed of light *(c)* squared. Nuclear power plants attempt to harness the tremendous amounts of energy unleashed by the disappearance of a tiny amount of matter.

The matter specifically selected and processed for this use in atomic power plants is called **nuclear fuel.** It is usually a compound of **uranium.** Uranium is made up of atoms that are enormous on an atomic scale.

Many elements emit radiation while they decay toward atomic stability. A few elements—types of uranium and plutonium, for example—**fission,** or split because of their instability. The fission most commonly used in U.S. nuclear reactors involves **uranium-235** (U-235). By absorbing a **slow,** or **thermal,** neutron, U-235 becomes **U-236**—a highly unstable atom that almost instantly fragments into two unequal sections, called **fission products.** Each U-236 breakup also releases an average of 2.3 neutrons, which are free to strike other U-235 atoms to make them undergo fission. When the number of neutrons released is sufficient to keep the fission process self-sustaining, a **chain reaction** is said to be in progress. Each fission releases energy because some of the matter of the nucleus disappears, and the two resulting fission products are more stable than the original atom they sprang from. They require less nuclear binding energy to hold their protons and neutrons in place, and the extra energy that was needed to hold the mammoth uranium nucleus together hurls the two fragments and the neutrons apart with tremendous force. The matter and nuclear binding energy are transformed into energy of atomic motion, or heat. This is the heat that ultimately boils the water to spin the electricity-generating turbines.

Fission takes place in a reactor's **core** and requires the presence of a **moderator**—something to moderate, or slow down, the neutrons so they will be traveling at the right speed for absorption by U-235. In most reactors, ordinary water acts as a moderator. Neutrons zipping between uranium atoms are slowed as they bump into the water molecules. These slowed **thermal neutrons** can keep the chain reaction going. Fast, unmoderated neutrons are almost useless as far as fission is concerned, but they are very important when it comes to making plutonium-239.

When a fast neutron is absorbed by a U-238 atom (and 97 percent of the uranium atoms in the core of a typical nuclear power reactor are of this variety), it creates U-239, which, after emitting two beta particles, becomes plutonium-239. **Fast reactors,** designed to maximize the use of fast neutrons, are built to manufacture plutonium for weapons use. In conventional nuclear plants, where slow neutrons are desired, about a quarter ton of plutonium is unavoidably created each year by neutrons traveling too fast to cause fission.

All reactors require a **coolant** to prevent the fission-heated nuclear fuel from melting and to conduct the heat away from the fuel and to the turbines. In most American power reactors, water doubles as both a coolant and a moderator.

The fission products, fragments of the original uranium nucleus, emit heat because of their intense radioactivity. The energy is given off as the newly formed atoms emit **alpha, beta, and gamma radiation.** This **decay heat** accounts for about one-tenth of the heat generated by a nuclear core. But it is this heat that creates the threat of a **meltdown** when a core is deprived of coolant, as was the case at Three Mile Island. The longer the fission process goes on, the more **hot** fission products are created. An old core releases much more decay heat, and would melt down much faster, than a young core. When it suffered its loss of coolant, TMI Unit Two's core was one of the "youngest" in the nation. Its relatively small amount of fission products helped keep its decay heat below the melting stage, lessening the danger posed by the accident. Heat generated by fission plays almost no role when a conventional reactor loses its coolant. Without coolant/moderator, fission stops.

Some of the more dangerous fission products—marked by their unusual atomic weights—include **iodine-131, cesium-137,** and **strontium-90.** All three travel up the food chain and can end up being concentrated in human tissues, emitting beta radiation to surrounding cells. Iodine-131, one of the main releases at Three Mile Island, concentrates in the thyroid gland and can cause cancer there decades after initial exposure. The body is fooled by strontium-90, mistaking it for calcium. This beta emitter becomes concentrated in bone tissue, where it can cause bone cancer or leukemia. The cesium isotope finds its way into muscle tissue. Many other fission products, including so-called **inert gases,** lodge in different parts of the body, silently inducing the formation of tumors and genetic defects. Some of these poisons are so long-lived that they act as permanent environmental contaminants, recycling through the ecosystem for thousands of generations. This is why plutonium, which remains lethal for more than a quarter million years, has been called "Thalidomide forever."

Virtually all of the hazards attributed to the use of nuclear energy to generate electricity share one common thread: they are unnatural. The deadly fission products discussed above don't exist in most natural systems. Plutonium is a man-made element. Uranium-235, the fissionable uranium needed for sustained chain reactions, accounts for less than 1 percent of the uranium found in nature. From beginning to end, the nuclear fuel cycle is an energy intensive, artificially contrived way to generate electricity. At each step of the cycle, the radioactivity of the substances handled increases, as does the corresponding threat to life. At each stage, workers, shippers, and handlers are exposed to potentially carcinogenic doses of radioactive poison. And the last stage of the cycle is yet to be developed. There is no known safe way to contain radioactive wastes for the eons it will take the fission products to lose their virulent radioactivity. More than one author has dubbed it "The Cycle of Death."

THE FRONT END OF
THE FUEL CYCLE

The Search for Fissionable Material: Mining

Uranium is a relatively scarce natural element usually found in sandstone deposits. High-grade uranium **ore** contains up to 3 percent uranium, and only a tiny fraction of that uranium is actually fissionable. As increasing amounts of uranium are mined for military and commercial uses, the higher grades of ore are being depleted. **Ores** with less than 1 percent uranium are now being excavated and processed. Uranium is found in the American West (where

Credit: U.S. Atomic Energy Commission.

THE NUCLEAR FUEL CYCLE

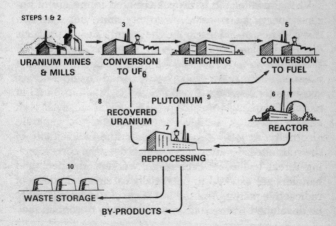

much of the ore lies under Native American reservations),
Canada, Australia, southern Africa, France, and the Soviet
Union.

Uranium miners have a long history of work-related lung
cancers and other diseases caused by prolonged exposure
to airborne radioactive contaminants. The hows and whys
of miner health problems are described in the article on
radiation's health effects.

Milling

The companies that mine uranium usually operate mills
within a few hundred yards of the excavation. Here the ore
is crushed and chemically processed into **yellowcake,** a
form of concentrated ore that is roughly 85 percent uranium compound. Yellowcake looks like its name. A ton of
raw ore will yield 4 pounds or less of yellowcake. As lower
grades of ore are used, this proportion will drop even
lower.

The leftovers from the milling process are millions of
tons of a powdery substance referred to as **tailings.** Mountains of these tailings rest near uranium mines and mills,
gently wafting ore wastes into the wind. While tailings
themselves are not highly radioactive, they contain **radium, radon,** and what physicists call **radon daughters**—all
of which are powerful alpha emitters that can cause lung
cancer once inhaled. Governments are only starting to regulate these wastes, which send unknown amounts of long-
lived carcinogens into the atmosphere.

Enrichment

Conventional nuclear reactors rely on U-235 to supply
heat-generating fissions to power their turbines. But only
0.7 percent of natural uranium is U-235. The rest is relatively nonfissionable U-238. To make chain reactions possi-

Uranium mill. Credit: Department of Energy.

ble, the proportion of fissionable uranium must be raised to at least 3 percent of the uranium volume. This **enrichment** of the fuel is a costly, energy-consuming process.

First, the yellowcake must be chemically converted into a gas, typically **uranium hexaflouride.** The uranium must be in a gaseous state for the actual enrichment process, which involves passing the gas through a special type of filter. U-235 is a little lighter than U-238, so it passes through the filter just a tiny bit faster than its nonfissionable isotope. The difference in their abilities to pass through the filter is so slight that the gas must be recirculated through the barrier more than a thousand times to bring the fissionable concentration up to the 3 percent necessary to maintain a chain reaction in a conventional reactor core. In the U.S., this process takes place at three enormous plants built with U.S. taxpayer funds. The plants, in Ohio, Kentucky, and Tennessee, consume about 3 percent of America's total electricity production.

Fuel Fabrication

After the uranium is enriched, it is chemically reconverted into a solid—typically uranium oxide—and formed into **pellets** designed to stack inside metal **fuel rods.** The long, slender rods are capped, sealed, and welded into **assemblies** that contain up to several hundred individual rods. The assemblies, which can now be called **fresh nuclear fuel,** are then shipped to a nuclear power plant for use. The core of a large reactor will contain several hundred assemblies weighing a hundred tons or more.

Nuclear Reactors

There are two main types of nuclear reactors: **water-cooled** and **gas-cooled.** Coolant of some type is necessary to draw off heat to spin a turbine, which makes electricity, and to prevent the fuel rods from melting. Most reactors use regular or **light water** as a coolant. In Canada, most reactors use rare **heavy water** for coolant. **Gas-cooled reactors** usually use helium or carbon dioxide. Every commercial reactor in the United States except one, the helium-cooled Fort St. Vrain plant in Colorado, uses light water, so **light-water reactors** (LWRs) are the focus of this book.

As if all these distinctions were not enough, there are two types of light-water reactor: **pressurized-water reactors** (PWRs) and **boiling-water reactors** (BWRs). The ill-fated reactor at Three Mile Island was a PWR built by Babcock and Wilcox. Other U.S. firms building PWRs include Westinghouse (which sells more reactors than any other reactor builder) and Combustion Engineering. The only BWR manufacturer is General Electric, which, at this writing, is rumored to be dropping out of the reactor market. About two-thirds of America's nuclear power plants rely on PWRs.

The Core of the Matter

After fuel assemblies arrive at a nuclear power plant, they are gingerly lowered into the reactor core. The assemblies are arranged according to a delicate geometry, one that will provide the proper conditions for a sustained chain reaction. At specific intervals, **control rods** slip into

Fuel rods being loaded into a reactor core. Credit: Iowa Electric Light & Power Co.

place. These rods contain special neutron-absorbing compounds that control the chain reaction by controlling the number of neutrons moving through the core. Usually, the rods are packed with **boron** or **cadmium.** By raising or lowering these rods, reactor operators try to regulate the intensity of the chain reaction. The control rods are also one of the reactor's prime lines of defense against a nuclear accident. Automatic nuclear-plant safety systems and, if necessary, manual controls are supposed to jam the control rods quickly all the way into the core in case of emergency. This is called a **scram** in the jargon of the nuclear industry and will usually stop the chain reaction in a normally operating core. Both the fuel and control rods are immersed in water, the coolant/moderator.

With fresh fuel and fully inserted control rods, the core is ready to start up. As the rods are slowly pulled out of the core in a pre-set pattern, more and more neutrons shoot between the fuel rods. The exteriors of the fuel rods (collectively called the **cladding**) do not block the neutrons. The reactor would never work if they did, because only neutrons passing between rods, and through the moderator, would be slowed sufficiently to force U-235 atoms to fission. The cladding, often fabricated from an alloy of **zirconium,** is not supposed to let the deadly fission products leach out into the coolant. The heat generated by the fission reaction in each rod is conducted through the cladding to the water, which cools the rods and transfers the heat to a turbine system. To keep the water moving, massive pumps are needed. And, for safety reasons, there are several backup water-circulating systems. But no matter how many backups there are, human and equipment errors can quickly let the coolant boil off, as happened at TMI.

A **Loss-of-Coolant Accident** (LOCA) would stop the fission reaction by depriving the neutrons of a moderator.

But the decay heat generated by the radioactive fission products would soon melt through the cladding and rupture the fuel rods, possibly leading to a meltdown. Without coolant, the decay heat would quickly buckle the fuel rods. The mass of twisted cladding and fuel pellets would soon start to melt. If coolant, often supplied by the **Emergency Core Cooling System** (ECCS), does not keep the core covered, it will melt down through the bottom of the reactor, and then burn its way through the floor of the **containment building,** and on, so to speak, toward China. This is where the eerily prescient movie *The China Syndrome* got its name. In reality, soil and ground water would slowly cool the core well before it completed its journey. Just how far it would get, no one knows.

A meltdown is just about the worst thing that can happen at a nuclear power plant. No one knows precisely what would happen if a core was to melt down, but here are some of the likely consequences: Most of the remaining coolant would boil into steam because of the intense heat. This sudden increase in pressure could lead to steam explosions that would rupture the reactor, releasing deadly fission products in the containment building. The explosions and heat could burst the containment building, leading to a **breach of containment** and shattering the reactor's last line of defense. Once the containment is breached, fission products are scattered by the wind—invisibly and silently.

A core melting through the floor of the containment building would contaminate the soil beneath the plant and any groundwater sources it encountered. A hot core hitting groundwater could trigger still more steam explosions, spewing radioactive soil, water, and steam into the atmosphere for wind dispersion. Radioactivity would also migrate through the soil and water supplies underground.

One U.S. government study, commissioned years before

the accident at Three Mile Island, concluded that an area the size of Pennsylvania would be contaminated following a meltdown.

The U.S. Nuclear Regulatory Commission, the federal agency ostensibly responsible for ensuring the safety of atomic power plants, has virtually ignored the catastrophic potential of meltdowns by basing its licensing process on so-called safety studies that deliberately discount the possibility and dangers of such mishaps.

During routine operation, the temperature at the center of the fuel rods can go up to 4000 degrees Fahrenheit. The entire point of a nuclear reactor is to put this tremendous heat to use. The specifics of how different types of reactors use this heat to generate electricity are detailed in "A Field Guide to Nuclear Power Plants" at the end of this article.

THE BACKED-UP END OF THE NUCLEAR FUEL CYCLE

Reprocessing

Among the countless assumptions that underpinned the early years of nuclear power development—assumptions that presumed safe, clean, and economical operation of atomic plants—was the assumption that **spent** (used-up) nuclear fuel would be **reprocessed**. Leftover fissionable uranium and plutonium would be ferreted out of the **spent fuel rods, refabricated** into new fuel rods, and reused in nuclear power plants. On paper, it's a very attractive idea. Nuclear proponents claimed 25 percent of the old fuel could be recycled. In practice, reprocessing has proven to be a commercial and political nightmare, highlighted by the economic collapse of facilities constructed to handle spent fuel from America's commercial nuclear power program. (The stillbirth of commercial reprocessing in the U.S.

is detailed in the article titled "Who Got Us into This Mess?")

Every year, one-quarter of the core of each nuclear power plant must be replaced because it has been used up. There are no longer enough fissionable U-235 atoms in those fuel rods to sustain a chain reaction. This leftover U-235 makes up about 0.8 percent of each rod. Plutonium comprises about 0.6 percent of each rod. The idea behind reprocessing is to separate out these two fissionable materials and recycle them so they can be used as fresh fuel at a later date.

After the spent fuel is taken from the core, it is immediately placed in a **spent-fuel storage pool** on the plant site. There, immersed in water that glows because of the intense radiation emitted by the rods, the spent fuel must cool for at least 120 days so that the shorter-lived fission products can decay and the fuel can be handled for shipment to a reprocessing facility. So far, so good. The problem is that there are no commercial reprocessing plants. With no outlet, utilities are rapidly filling their spent fuel storage pools to capacity—and once a pool is full, the reactor must shut down. Right now there is no other place to put the used fuel.

A dozen or more U.S. reactors face the prospect of such a forced shutdown within the next decade. This has sent many utilities into a mild panic, and the Nuclear Regulatory Commission has started to issue orders authorizing **compaction** of spent fuel in the pools. Compaction is a delicate way of saying that the federal government is allowing utilities to stuff more lethally hot radioactive fuel in its storage pools than those pools are designed to handle.

Reprocessing plants are currently being operated in France, England, Japan, and the Soviet Union. After spent fuel arrives at the facility, it is put in yet another storage pool for further cooling. Then the rods go into a vat of boiling nitric acid where the contents of the fuel rods dissolve. Fissionable materials are separated and extracted chemically. They are returned to the front end of the fuel cycle for reconversion into "fresh" fuel. The acidic sludge left behind is one of the more exotic species of high-level radioactive waste.

Waste: No Answers

The problem with rapidly filling on-site storage pools illustrates the entire dilemma surrounding nuclear waste disposal. The deadly waste keeps piling up, and there is literally no place to put it. Just before the core melt at Three Mile Island, the umpteenth nuclear-waste-disposal task force assembled by the federal government issued a report concluding that no proven means of handling the wastes (which remain toxic for thousands of *generations*) exists, and that still more research needs to be conducted. One of the major fallacies propagated by the nuclear industry is that the technical means for disposing of radioactive waste are at hand and that Washington politics is the culprit behind the radwaste scandal. It's true that the federal nuclear bureaucracies have been analyzing and reanalyzing the nuclear-waste problem for decades. But that does not alter the fact that no one knows what to do with the stuff.

Even if some solution could be found for isolating contaminated clothing, highly radioactive sludge, and spent fuel, the question of what to do with worn-out nuclear power plants remains. Nuclear power plants have a life span of 30 to 40 years. After that, the plant's components are so radioactive, and so clogged with lethal crud, that they must be shut down. Congressional researchers estimate that parts of the closed, or **decommissioned,** atomic plants will remain dangerous for more than one million years. The problems posed by radioactive waste are detailed in another article later in this chapter.

Was Murphy an Engineer?

In countless laboratories, factories, and offices, a quote is
conspicuously posted. "Murphy's Law: Whatever Can Go
Wrong, Will Go Wrong." Murphy should have been a nuclear engineer. Like any large system where people and
complex machinery mesh, the nuclear-fuel cycle falls victim to mistakes, be they human or mechanical. A fuel rod
gets stuck. A reactor component breaks loose and obstructs
coolant flow starting a core melt. Water piping systems get
cross-connected and plant drinking fountains are suddenly
linked to radioactive waste tanks. All these things have
happened, and more foul-ups take place every day at nuclear power plants. Hundreds are recorded by the NRC
every month. Whatever can go wrong, does.

Murphy's unshakable law creates special problems for
nuclear power. With tons of toxic solids, liquids, sludges
and gases, the nuclear fuel cycle is unforgiving. A full-size
nuclear power plant is home to more radioactive material
than 1000 Hiroshima-sized atomic bombs. One miscalculation, however small, or one act of sabotage, could cause
tens of thousands of deaths and billions of dollars in property damage.

How close have we come to atomic-power holocausts?
Well, one nuclear engineer quipped, "We almost lost Detroit" after a partial core melt at an experimental breeder
reactor just outside the motor city in October of 1966. New
York City might become a radioactive desert if a core at
the finicky Indian Point nuclear plant 24 miles north of the
city melts down. We may lose some of the world's most
fertile farmlands should an earthquake strike one of the
atomic plants sited near a geologic fault in California. And
it may take years before we know how close we came to
losing Harrisburg, Pennsylvania.

Nuclear advocates (and many of these are engineers)

love to talk about their "defense in depth" approach to the safety of nuclear power plants. Redundant safety systems are said to be layered upon still more redundant safety systems until the **probability** (read odds) of an atomic mishap become so small that it is worth the risk. This is the myth TMI shattered.

What good are redundant safety systems if a plant maintenance person leaves backup feedwater loop valves sealed? How can even the best-trained nuclear-plant operators regulate the level of coolant in the reactor when there is no gauge to tell them what the level actually is? How can the Emergency Core Cooling System supply the coolant needed to keep a core from melting down when the people running the plant turn the ECCS off, in response to misleading indicators that show the core is already being adequately cooled? As this book goes to the printer, the Nuclear Regulatory Commission staff is trying to answer these questions. But they are still allowing several dozen nuclear plants around the United States to continue "normal" operations.

No one knows how safe (or how dangerous) nuclear plants are. There is no official study that conclusively outlines the risks associated with nuclear power. The Rasmussen Report—long hailed by nuclear power's boosters because it claimed the odds of a major nuclear accident were similar to the chances of one's being struck by a meteor—was disavowed by the NRC two months before TMI. A joke circulating in the days following TMI claimed that Dr. Norman Rasmussen, a professor at the Massachusetts Institute of Technology who piloted the discredited study, had been felled by a rock that fell from the sky.

One reason hard information about nuclear safety is so difficult to come by is that construction of bigger and bigger nuclear plants raced ahead of the research that was being done on the technology's reliability. A full-scale Emergency Core Cooling System test on a potentially melting core has never been performed. Information used to design the ECCS was based on vast extrapolations of computer codes that have never been accurate in predicting how a reactor would behave during an accident.

But a reactor need not suffer a Loss-of-Coolant Accident caused by a broken pipe (until TMI, the most widely discussed type of potential snafu) in order to threaten the public. Day-to-day operation of a plant involves routine **low-level radiation** emissions that add to the cancer risk of people living downstream and downwind. Current research indicates that no level of radiation exposure is truly safe. This is why atomic energy has no room for error.

Nuclear power is indefensible on technical grounds. Millions of tons of the most toxic substances known to man are handled by a capricious, loosely regulated technology whose behavior cannot be predicted even by the people who created it. Murphy would not approve.

A FIELD GUIDE TO NUCLEAR POWER PLANTS

How many ways are there to boil water? The world's top nuclear scientists have been puzzling over this question for decades, and have come up with five broad answers. All involve using tons of radioactive fuel, a fine-tuned nuclear chain reaction, mind-boggling webs of piping, and a seemingly endless litany of redundant backup safety systems to protect us from the fuel that the scientists have gone through so much trouble to create.

This is how the five different types of reactors work.

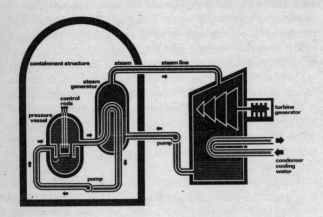

Pressurized water reactor (PWR)

Atomic Industrial Forum, Inc.

Pressurized-Water Reactors

In a **pressurized-water reactor** (PWR) like Three Mile Island, the reactor core sits in a tightly sealed pressure vessel. The entire **primary coolant system,** which includes the core, its coolant, and the huge pumps that are supposed to keep the water flowing, is kept at abnormally high pressure—over 2000 pounds per square inch—in order to prevent the superheated water from boiling. Pressure in the system is regulated by the **pressurizer,** which holds a mixture of steam and water linked to the primary coolant system and is regulated by the reactor operators. By raising or lowering the temperature of the pressurizer (and thus changing the ratio of water to steam in that cylinder), operators in the plant's **control room** can, to a degree, control the pressure of the primary coolant system.

At Three Mile Island (TMI), a valve at the top of the pressurizer stuck open and leaked primary coolant all over the floor of the containment building. The operators did not know about the stuck valve, and thought they were controlling the pressure just fine.

Assuming the primary coolant system is operating as it is supposed to, the coolant will draw heat away from the core and pump it to a **steam generator.** A steam generator is simply a device by which the primary coolant system is brought close to the **secondary coolant system.** In Westinghouse reactors, the primary fluid, bearing heat from the nuclear core, circulates through thousands of thin U-shaped tubes that are enveloped by cooler, secondary coolant. In Babcock and Wilcox plants, like TMI, the tubes are straight. For all the effort expended to generate tremendous concentrated heat in the reactor's core, a relatively small amount of heat is passed through the steam generator to the turbines. At the crippled Pennsylvania reactor's twin, TMI Unit One, primary cooling water leaves the core and enters the steam generator (through the so-called **hot leg** pipes) at 604 degrees Fahrenheit. It loses a mere 50 degrees to the secondary coolant and returns to the core (via the **cold leg**) at 554 degrees. The secondary coolant, which is not kept under extremely high pressure, boils into steam as it passes around the primary coolant tubes.

Now in vapor form, secondary coolant is channeled to the **turbines,** which spin as the steam rushes against their pinwheel-like blades. Now a third cooling system comes into play. Water from outside the nuclear station, from a lake, a river, or the ocean, is sucked into the plant to cool the secondary coolant in the **condenser.** The cooled-off secondary fluid condenses from steam into liquid water, and the newly heated water from outside the plant is (after a minimum amount of cooling) returned to the environment somewhat warmer than it was when it left. This is called **thermal pollution,** and has been blamed for many fish kills and ecological disruptions. Finely tuned ecosystems frequently cannot withstand the sudden shock of even a small temperature change.

The secondary coolant, back in liquid form, is sent by **feedwater pumps** back to the steam generator for another go-round.

Boiling-Water Reactors

Boiling-water reactors (BWRs) are less complex than PWRs, because they have only two coolant loops. The reason they are called boiling-water reactors is that the primary coolant is allowed to boil after it comes in contact with the core. The reactor thus bypasses the need for a steam generator or a secondary coolant system. The steam spins the turbines and is condensed by water pumped in from outside the plant. The primary coolant is then sent,

Boiling water reactor (BWR)

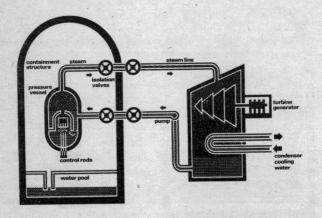

Atomic Industrial Forum, Inc.

via feedwater pumps, back to the core to reboil into steam and repeat the cycle.

Although this type of reactor seems simpler than the PWR, it is no less dangerous. Because the same water circulates between the radioactive core and the turbines, much more equipment becomes contaminated. In a PWR, in theory, all the radioactivity is contained within the primary loop. With BWRs, virtually every piece of equipment involved in turning the core's heat into electricity becomes, in the radioactive sense, **hot**.

High-Temperature Gas-Cooled Reactors

The Fort St. Vrain reactor, the only non-light-water power reactor in the United States, uses helium the same way a PWR uses primary coolant: to boil water circulating through a steam generator. It then uses that steam to spin

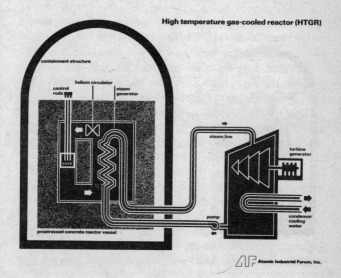

High temperature gas-cooled reactor (HTGR)

containment structure

helium circulator

control rods

steam generator

core

steam line

turbine generator

prestressed concrete reactor vessel

pump

condenser cooling water

⚠F Atomic Industrial Forum, Inc.

turbines. Helium, though a relatively effective coolant, is a poor moderator, so this reactor uses a carbon compound called **graphite** to moderate the chain reaction. Several European nations, particularly Great Britain and France, started developing such **high-temperature gas-cooled reactors** (HTGRs) in the 1950s, only to find themselves outmaneuvered politically and economically by the light-water-reactor forces in the United States.

Heavy-Water Reactors

CANDU, shorthand for **Canadian deuterium uranium reactor,** is the upbeat name for Canada's unique reactor—the only one explicitly developed for peaceful uses and, ironically, the type of reactor first used by India to make plutonium for its illicit weapons program. The CANDU uses **heavy water** for its moderator/coolant. The difference between light and heavy water lies in the hydrogen nuclei of the two different types of water molecules. Ordinary, or light, water has two regular hydrogen atoms bonded to one oxygen atom—the familiar H_2O. Deuterium is different. Its hydrogen atoms are twice as heavy as those found in ordinary water. Instead of having only a single proton for a nucleus, the deuterium atoms contain a proton and a neutron; hence the name *heavy water.*

Heavy water occurs naturally but in extremely small amounts—something like the relatively small percentage of uranium ore that is fissionable. As such, the production of heavy water requires a great deal of effort. Why bother? Because it saves the trouble of enriching the uranium. Heavy water is such an efficient moderator that CANDUs are the only reactors in the world that use natural, unenriched uranium for fuel. The heavy water's extraordinary moderating capabilities compensate for the low volume of fissionable uranium found in nature. Mechanically, CANDUs function somewhat like PWRs, with the heavy water

acting as primary coolant to boil water in a secondary coolant system in a steam generator.

Liquid-Metal Fast-Breeder Reactors

Finally, there is the highly touted **breeder reactor**—the
nuclear technology that can produce more fuel than it
consumes. It literally breeds plutonium. It can also do
something else no other atomic reactor can do: produce
nuclear explosions identical to the type used in nuclear
weapons.

Most breeders fall into the category of **liquid-metal fast-
breeder reactors** (LMFBR). LMFBRs use **liquid sodium** as a
coolant because of its potent heat-transferring characteristics and because it is *not* an efficient neutron moderator.
Without effective moderation, the neutrons in LMFBR
cores remain fast—and are able to transform nonfissionable
uranium-238 into fissionable plutonium-239. (This is dis-

Liquid metal fast breeder reactor (LMFBR)

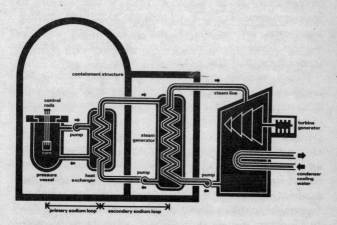

Atomic Industrial Forum, Inc.

cussed in detail in the Gone Fission section of the preceding article.) Breeder reactor cores are encased with a blanket of U-238 to capture the fast neutrons and breed plutonium for use as reactor fuel or explosives. In theory, a breeder can produce beyond 40 percent more plutonium than it uses. By combining both power production and fuel production chores in one reactor, LMFBRs are supposed to stretch the world's uranium supplies and make nuclear generation of electricity possible for thousands of years.

That scenario, fortunately, is unlikely. Breeder technology has become one of the most punted political footballs in Washington, with billions of dollars spent on a program that has not generated a single commercial kilowatt of electricity. But opposition to the breeder rests on technical as well as economic grounds.

Where BWRs use two cooling systems and PWRs use three, breeders use four. To make matters worse, the primary (core cooling) and secondary (initial heat transfer) loops are filled with pressurized liquid sodium—an element that burns on contact with air or water. This means a leak in either of the first two cooling systems could lead to a chemical explosion in the nuclear plant. Such an explosion could easily unleash deadly nuclear fission products. Sodium could also react with the concrete of a containment building to release explosive hydrogen gas.

Another special characteristic of the breeder is that it relies on plutonium as its primary fuel. Fast reactors require a smaller, more densely packed core that is more vulnerable to a small disruption in coolant flow. If coolant is blocked, a small number of fuel rods melting together could provide enough fissionable material (a **critical mass**) for a low-grade nuclear explosion—one that could rupture the containment building, ignite the sodium coolant, and spew plutonium and its fission products into the air.

LMFBRs are functionally similar to PWRs. The pressurized sodium courses through the core and the primary

coolant loop, giving up some of its heat to a secondary
sodium cooling loop. This second loop then surrenders its
heat in a steam generator that is incorporated into a third
cooling loop, which contains regular water. This water
boils into steam, spins turbines, and is condensed by cooling water from a fourth, usually natural, cooling water
source before being recirculated to the steam generator by
feedwater pumps.

There are two safety-related reasons for the addition of
the secondary sodium loop. By having the steam generator
one step removed from the radioactive core-heated primary sodium loop, the danger from the inevitable steam
generator leaks is lessened. Ordinary nuclear-reactor
steam generators are notorious for developing leaks between the two cooling systems. Such a leak in an LMFBR
could be disasterous because it could instantly touch off a
chemical explosion when the sodium and water meet. If
LMFBRs had primary coolant flowing through the steam
generator, such an explosion would instantly spread radioactivity and possibly lead to a nuclear explosion by disrupting the core cooling system. The other reason for a
second sodium loop is to create a smoother heat transfer
system between the core-heated sodium and the light
water.

Despite myriad built-in technical handicaps (as well as
the problems created by manufacturing enormous quantities of plutonium), France and the Soviet Union have made
major economic commitments to develop breeder reactors
as a mainstay of their energy-supply systems. The breeder
is also the main U.S. government-funded energy research
project.

—G.R.Z.

The Browns
Ferry Incident

David Dinsmore Comey

At noon on March 22, 1975, both Units One and Two at the
Browns Ferry plant in Alabama were operating at full
power, delivering 2200 megawatts of electricity to the
Tennessee Valley Authority (TVA).

Just below the plant's control room two electricians were
trying to seal air leaks in the cable spreading room, where
the electrical cables that control the two reactors are separated and routed through different tunnels to the reactor
buildings. They were using strips of spongy foam rubber to
seal the leaks. They were also using candles to determine
whether or not the leaks had been successfully plugged—
by observing how the flame was affected by escaping air.

The electrical inspector put the candle close to the foam
rubber, and it burst into flame.

The resulting fire, which disabled a large number of the
engineered safety features at the plant, including the entire Emergency Core Cooling System on Unit One and
almost resulted in a boiloff/meltdown accident, demonstrates the vulnerability of nuclear plants to "single failure" events and human fallibility.

The fire was started by an electrical inspector (referred

to in the Nuclear Regulatory Commission's [NRC] report as "C") working with an electrician, "D," who said,

> Because the wall is about 30 inches thick and the opening deep, I could not reach in far enough, so C [the inspector] asked me for the foam and he stuffed it into the hole. The foam is in sheet form, it is a "plastic" about two inches thick, that we use as a backing material.

The inspector, "C," described what happened next:

> We found a two-by-four-inch opening in a penetration window in a tray with three or four cables going through it. The candle flame was pulled out horizontal showing a strong draft. D [the electrician] tore off two pieces of foam sheet for packing into the hole. I re-checked the hole with the candle. The draft sucked the flame into the hole and ignited the foam, which started to smolder and glow. D handed me his flashlight, with which I tried to knock out the fire. This did not work and then I tried to smother the fire with rags stuffed in the hole. This also did not work and we removed the rags. Someone passed me a CO_2 extinguisher with a horn, which blew right through the hole without putting out the fire, which had gotten back into the wall. I then used a dry chemical extinguisher and then another, neither of which put out the fire.

In its report on the cause of the fire, the TVA stated:

> The material ignited by the candle flame was resilient polyurethane foam. Once the foam was ignited, the flame spread very rapidly. After the first application of the CO_2, the fire had spread to the reactor building side of the penetration. Once ignited, the resilient polyurethane foam splattered as it burned. After the second extinguisher was applied, there was a roaring sound from the fire and a blowtorch effect

due to the airflow through the penetration.

The airflow through the penetration pulled the material from discharging fire extinguishers through the penetration into the reactor building. Dry chemicals would extinguish the flames, but the flame would start back up.

Approximately 15 minutes passed between the time the fire started (12:20 P.M.) and the time at which a fire alarm was turned in. It was not until one of the electricians told a plant guard inside the turbine building that a fire had broken out that an alarm was sounded. However, confusion over the correct telephone number for the fire alarm delayed its being sounded.

As the NRC report on the incident noted:

> The Browns Ferry Nuclear Plant Emergency Procedure lists two different telephone numbers to be used in reporting a fire, one in a table of emergency numbers and the second in the test of procedure. The appropriate number (299) is the one in the test; dialing this number automatically sounds the fire alarm and rings the Unit One operator's telephone.
>
> The Emergency Procedure was not followed by those involved when [they] reported the fire. The construction workers first attempted to extinguish the fire, whereas the procedure specifies that the fire alarm be sounded first. The guard reporting the fire telephoned the shift engineer's office rather than calling either of the numbers listed in the procedure.

Only when the shift engineer then called the control room on the 299 number to get the reactor operator was the plant fire alarm actually sounded. It was fortunate that the shift engineer was in an office with a PAX phone (the plant's internal telephone system), which allowed him to call the 299 number. Had he been at a construction department extension, he could not have placed the call, as the TVA's investigative report later revealed:

> BFNP Standard Practice BFS3, "Fire Protection and Prevention," instructs DPP (Department of Power Production) personnel discovering a fire, whether in a construction area or an area for which DPP is responsible, to report the fire to the Construction Fire Department, telephone 235. "BFNP Fire, Explosion and Natural Disaster Plan" instructs personnel discovering a fire to call 299 (PAX). The construction extension cannot be dialed from the PAX system, and the plant extension cannot be dialed from the Construction phone system.

Despite the fire alarm, the reactor operators in the plant control room did not shut down the two reactors, but continued to let them run. At 12:40, five minutes after the fire alarm sounded, the Unit One reactor operator noticed that all of the pumps in the Emergency Core Cooling System (ECCS) had started. In addition, according to the official TVA report,

> Control board indicating lights were randomly glowing brightly, dimming, and going out; numerous alarms occurring; and smoke coming from beneath panel 9-3, which is the control panel for the emergency cooling system. The operator shut down equipment that he determined was not needed, only to have them restart again.

The flashing lights, alarms, smoke, and continual restarting of ECCS pumps went on for a full ten minutes before the reactor operators began to wonder whether it might be prudent to shut down the reactors.

After the power level of the Unit One reactor began to drop inexplicably, the operator started to reduce the flow of the reactor's recirculating pumps; when the pumps sud-

denly quit at 12:51, he finally shut the reactor down by
inserting the control rods.

Beginning at 12:55, the electrical supply was lost both to
control and to power the Emergency Core Cooling System
and other reactor shutdown equipment on Unit One. The
normal feedwater system was lost; the high-pressure ECCS
was lost; the reactor-core spray system was lost; the low-
pressure ECCS was lost; the reactor-core isolation cooling
system was lost; and most of the instrumentation that tells
the control room what is going on in the reactor was lost.
According to the Unit One operator,

> I checked and found that the only water supply to the
> reactor at this time was the control rod drive pump, so
> I increased its output to maximum.

Meanwhile, a few feet away, on the Unit Two side of the
control room, warning lights had also been going off for
some time. A shift engineer stated:

> Panel lights were changing color, going on and off. I
> noticed the annunciators on all four diesel generator
> control circuits showed ground alarms. I notified the
> shift engineer of this condition and said I didn't think
> they would start.

According to the official TVA report:

> At 1 P.M., the Unit Two operator observed decreasing
> reactor power, many scram alarms, and the loss of some
> indicating lights. The operator put the reactor in shut-
> down mode.

Some of the shutdown equipment began failing on Unit
Two, and the high-pressure ECCS was lost at 1:45 P.M.
Control over the reactor relief valves was lost at 1:20 P.M.
and not restored until 2:15 P.M., at which time the reactor
was depressurized by using the relief valves and brought
under control.

On the Unit One side of the control room things were not going so well. According to the Unit One operator:

> At about 1:15, I lost my nuclear instrumentation. I only had control of four relief valves. . . .
>
> At about 1:30, I knew that the reactor water level could not be maintained, and I was concerned about uncovering the core.

Had the core become uncovered, a meltdown of the reactor fuel would have begun because of the radioactive decay heat in the fuel.

In order to prevent the reactor water from boiling off, it was necessary to get more water into the core than the single high-pressure control-rod drive pump could provide. It was decided that by opening the reactor relief valves, the reactor would be depressurized from 1020 to below 350 pounds per square inch, where a low-pressure pump would be capable of forcing water in to keep the core covered. None of the normal or emergency low-pressure pumps was working, however, so a makeshift arrangement was made, using a condensate booster pump. This was able to provide a temporarily adequate supply of water to the reactor, although the level dropped from its normal 200 inches above the core to only 48 inches. Using the makeshift system, the Unit One reactor was under control for the time being.

Unit Two was also under control, but by a rather thin margin. The "A" and "C" subsystems of the low-pressure ECCS and the core spray system had been lost early in the incident, and the "B" system failed intermittently between 1:35 and 4:35. With only one subsystem of the low-pressure ECCS available, the Unit Two operator resorted to using the condensate booster pump arrangement similar to the one that had been rigged up for Unit One.

Many instrumentation and warning lights in the control room were inoperative. The reactor protection system and nuclear instrumentation on both reactors had been lost shortly after they were shut down. Most of the reactor water level indicators were not working. The control-rod position indicator system was not operative. The process computer on Unit One was lost at 1:21 P.M. (The computer on Unit Two was inoperative because it was down for reprogramming.)

Other systems were failing; at 2:43 one of the plant's four diesel generators failed, leaving the plant with a bare minimum of emergency on-site power supply.

To add to the confusion, the PAX telephone system failed at 1:57 P.M., making outgoing calls from the control room impossible for several hours. This represented a considerable hardship, because the control room had lost control over most of the plant's valves, and the plant telephone system was being used to instruct equipment operators to manually adjust certain key valves in the condensate booster system pumping water into the reactor core.

Moreover, the Unit One operator did not know the level or the temperature of the water in the torus (the reactor containment suppression chamber) because the monitors were not working. Yet, as a General Electric supervisor's log showed,

> With the relief valves in operation, the need for torus cooling became vital. The RHR [residual heat removal] system was unavailable for torus cooling.

Unless the RHR system could be put into operation, there was the danger that the water in the torus would begin to boil, and this would eventually overpressurize the containment and rupture it.

And the NRC report remarks in its restrained way,

> After condensate flow to the reactor was established, the major concern was to establish torus cooling and shutdown cooling using the RHR as quickly as possible.

From about 2 P.M. until the fire was extinguished, several attempts were made to enter the reactor building and manually align the RHR for torus cooling and shut down cooling modes. . . .

None of these attempts resulted in establishing torus or reactor shutdown cooling. The attempts were severely limited by dense smoke and inadequate breathing apparatus.

The fire-fighting effort was not going well. Soon after the electricians had fled the cable spreader room, a shift engineer had tried to turn on the built-in Cardox system in order to flood the room with carbon dioxide and put out the fire. He discovered that the electricians had purposely disabled the electrical system that initiated the Cardox.

I tried to use the manual crank system and discovered that it had a metal construction plate on under the glass and I tried to remove it. This was difficult without a screwdriver. . . . The next day, I checked other manual Cardox initiators and found that almost all of them had these construction plates attached.

He finally got the power on, but the Cardox system ended up driving smoke up into the control room above the cable spreader room. One person present described the scene in the control room as follows:

The control room was filling with thick smoke and fumes. The shift engineer and others were choking and coughing on the smoke. It was obvious the control room would have to be evacuated in a very short time unless ventilation was provided.

After the carbon dioxide system was turned off, the smoke stopped pouring into the control room. It had not put out the fire in the spreading room, however. A safety officer fighting the fire pointed out:

The CO_2 in the spreader room may have slowed down the fire but did not put it out. We opened the doors for air, as the smoke in the whole area had become dense and sickening. Another employee and I each donned a breathing apparatus and went into the spreader room. We used hand lamps for illumination, but they penetrated the smoke only a few inches. The neoprene covers on the cables were burning, giving off dense black smoke and sickening fumes. . . . It was impossible not to swallow some smoke. I got sick several times.

Because of the close quarters in the spreader room, fighting the fire was difficult. One safety officer said:

I went into the spreader room wearing a Scott air pack and mask and carrying a fire extinguisher. I had to crawl under the cable trays. The air pack cylinder was too cumbersome to wear on my back so I took it off and slid it and the fire extinguisher under the trays about 30 feet to the fire.

Inoperative equipment also hampered the fire-fighting effort. For example, one assistant shift engineer said:

I returned to the spreader room to direct the fire-fighting effort. A wheeled dry chemical extinguisher had been brought to the spreader room, but its nozzle was broken off at the bottle and I told some of the men to get it out of there and find another unit.

The official Nuclear Regulatory Commission report noted other deficiencies:

Breathing apparatus was in short supply and not all of the Scott air packs were serviceable. Some did not have face masks and others were not fully charged at the time of the start of the fire. The breathing apparatus was recharged from precharged bulk cylinders by pressure equalization. As the pressure in the bulk cylinders decreased, the resulting pressure decrease in the Scott packs limited the length of the time that the personnel could remain at the scene of the fire.

One of the assistant unit operators who was sent into the reactor building to manually open the RHR cooling valves reported:

> We made three tries but could not get to the valves. Our breathing equipment could only supply 18 minutes of air per tank, which was not sufficient to enable us to get to the valves and back out of the area. The air tanks were being recharged, but the pressure in the main tanks was not strong enough to fill the tanks to their normal air supply. After the third attempt we went back to the control room and told the assistant shift engineer of the problem and that we needed different equipment or fully charged tanks to succeed.

The electrical cables continued to burn for another six hours, because the fire fighting was carried out by plant employees, despite the fact that professional firemen from the Athens, Alabama, fire department had been on the scene since 1:30 P.M. As the Athens fire chief pointed out:

> I was aware that my effort was in support of, and under the direction of Browns Ferry plant personnel, but I did recommend, after I saw the fire in the cable spreading room, to put water on it. The Plant Superintendent was not receptive to my ideas.
> I informed him this was not an electrical fire and that water could and should be used because the CO_2 and dry chemical were not capable of providing the required cooling. Throughout the afternoon, I continued to recommend the use of water to the Plant Superinten-

dent. He consulted with people over the phone, but
apparently was told to continue to use CO_2 and dry
chemical. Around 6 P.M., I again suggested the use of
water. . . . The Plant Superintendent finally agreed and
his men put out the fire in about 20 minutes.

They were using type B and C extinguishers on a type
A fire; the use of water would have immediately put the
fire out.

Even when the decision to put the fire out with water
had been made, further difficulties developed. The fire
hose had not been completely removed from the hose
rack, so that full water pressure did not reach the nozzle.
The fire fighters did not know this, however, and decided
that the nozzle was defective. They borrowed a nozzle
from the Athens fire department, "but it had incorrect
type threads and would not stay on the hose."

Once the fire was put out, it was possible for plant employees to go into the reactor building and manually open
valves to get the RHR system operating.

On Unit One, however, a new emergency developed.
About 6 P.M., control of the last four relief valves was lost,
and the reactor pressure increased to above 350 pounds
per square inch, making it impossible for the makeshift
condensate booster pump system to inject water into the
reactor. As in the early stage of the accident, the only
source of water for the Unit One reactor was now the
control-rod drive pump, and this probably would not prevent a boiloff accident that would turn into a core meltdown in just a few hours.

The spare control-rod drive pump was inoperative, and
although it was later determined that a series of valves
could have been turned to allow the Unit Two control-rod
drive pump to supply water for the Unit One reactor, the
reactor operators did not know this at the time.

With the reactor pressure mounting higher and higher,

the relief valves were finally brought back into operation at 9:50 P.M., and at about 10:20 P.M. the reactor was depressurized to the point that the condensate booster pump could again get water into the reactor.

Normal shutdown was established on the Unit One reactor at 4:10 the next morning. The nightmare at Browns Ferry was over.

Had the reactor boiloff continued to the point where a core meltdown took place, however, it is doubtful that the endangered surrounding population could have been evacuated in time; evacuation of the county's residents was the responsibility of the Civil Defense Coordinator for Limestone County, but, as he admitted to NRC inspectors:

TVA's Browns Ferry Nuclear Power Plant.

I heard about the fire at Browns Ferry on the morning of Monday, March 24, 1975 [two days later]. No one in the Civil Defense System notified me or attempted to do so. . . . I feel that our county should have been notified since the plant is located in our county.

The sheriff of Limestone County said:

I heard about the fire at the Browns Ferry plant after it was over. . . . I have not had any updating of procedures proposed to me since the initial plan was outlined in 1972. I do not have a copy of the emergency plan.

The sheriff of neighboring Morgan County did hear about the fire four hours after it started but said, "I was asked to keep quiet about the incident to avoid any panic." The NRC noted in its investigative report:

No official notification was made to the State of Alabama Highway Patrol by the State of Alabama Department of Public Health or by TVA. . . .

An attempt was made to notify the Lawrence County sheriff at 4:08 P.M. but no answer was received. Only one attempt was made to locate the sheriff.

In fact, this try-once-and-fail procedure was more or less the norm. The NRC investigation noted:

The State of Alabama Emergency Plan for the Browns Ferry Nuclear Plant was implemented at 3:30 P.M. to the extent that notifications were made to designated state personnel and principal support agencies. . . . Only one attempt was made to contact principal support agencies that were located in counties surrounding the site regardless of whether the agency was contacted or not. The notification process was discontinued at 4:40 P.M.

The [NRC] investigators commented to the Director

of Radiological Health that, due to the uncertainty relating to the status of the reactors from 12:30 P.M. to 7:45 P.M., the implementation of the state plan indicated that a "standby" classification was necessary that would have required continuous notifications and recommendations to be made to support agencies until the reactor was verified to be in a safe condition.

Additionally, some agency officials related that they did not have a copy of the state plan or the plan that they had needed updating. Other officials indicated that they had received very little information concerning their defined responsibilities relating to an emergency at the plant. . . .

The State of Alabama and BFNP personnel have participated in emergency drills to test the effectiveness of their emergency plans for the past several years. Participation in the drills by the state has involved the verification of notification procedures and the time required to travel to the site to perform environmental sampling.

The fire knocked out the radiation monitors on the Unit One reactor building vent almost immediately, and the Unit Two vent monitor was inoperable from about 2 P.M. until 9 P.M. Both the NRC and TVA state unequivocally that no significant radiation release occurred, but there were continuing difficulties in obtaining air samples both at the plant site and in the surrounding area. For example,

At 5:05 P.M., the Director, Environs Emergency Center, directed that additional environmental air samples be obtained. . . . At this time individuals in the Site Emergency Center observed smoke emanating from the reactor building and the decision was made to evacuate the meteorological tower.

Radiation sampling of air was started at 4:45 P.M. at Athens, 10 miles northeast of the plant; at Hillsboro, 10 miles southwest; in Rogersville, 35 miles northwest; but not at Decatur, 20 miles southeast and directly downwind of the plant.

The sampler at Decatur, Alabama, was thought to be inoperable possibly due to the wind direction control system, but the Laboratory Director was asked to investigate the problem. The Laboratory Director reported 7:50 P.M. that no air sampler was available at Decatur. This station would have been the major air station of importance because Decatur, Alabama, is located in the southeast direction from the site and the wind direction at the time of the fire was from the northwest section. Arrangements were made with the State of Alabama Air Pollution Control Commission for using one of their samplers at the Decatur station. Air sampling was initiated at this station at approximately 9 P.M., CDT, on March 22, 1975.

Other equipment failures also continued to plague the plant. Shortly after nightfall, the aircraft warning lights on the plant's radioactive gas release stack went out. Since the stack is 600 feet tall, loss of the light could have resulted in an aircraft colliding with the stack. The NRC report describes what was done next.

At 8:37 P.M. a member of the environment staff made an attempt to telephone the gatehouse by using a public telephone to inform the security guards that the warning lights on the plant stack were not operating. Since the gatehouse could not be reached, the environmental representative telephoned the EEC [Environs Emergency Center] and explained the condition. The Director, EEC, directed the information to the plant because of the need to contact FAA [Federal Aviation Administration] authorities immediately.

It is unclear why no one thought to phone the FAA directly instead of giving the information to the plant, which had more than enough problems on its hands at the time.

Other gremlins that cropped up included the plant's electric sequence printer running out of tape at 4:30 on the afternoon of the fire, so that information on the time and sequence of restoration of control circuits after that time was lost, since no one replaced the tape until 2 P.M. the following day.

At about 3:40 P.M., the decision was made to begin tape-recording all telephone communications between the plant and the chief of TVA's nuclear generation branch. But as the NRC report noted, "A review of the tapes revealed mechanical problems with the tape recorders and only partial transcripts were obtained."

Some of what was recorded, however, is indicative of the thought process of TVA and NRC personnel. One example is the following excerpt from a conversation at 7:47 P.M. between J. R. Calhoun, Chief of TVA's Nuclear Generation Branch, and H. J. Green at the Browns Ferry plant:

> GREEN: I got a call that Sullivan, Little and some other NRC inspector are traveling tonight and will get here sometime tonight and so all our problems will be over.
>
> CALHOUN *(laughs):* They will square you away, I am sure.
>
> GREEN: We probably have a violation. We've kept very poor logs.
>
> CALHOUN (laughs): No doubt!

At about 9 P.M., Calhoun phoned Frank Long, in the U.S. Nuclear Regulatory Commission's Region II office in Atlanta:

> LONG: The doggone public news media types will probably drive you out of your mind. Okay, your

people did put out a news press release?

CALHOUN: Yeh, we put one out about 4:30. Somewhere close to 4:30. . . . Only thing we can say right now is that it could have been a hell of a lot worse.

LONG: Oh yeah.

CALHOUN: You know, when you talk about a fire in the spreading room, you've really got problems.

LONG: It would affect just about everything.

CALHOUN: Yeah, you know everything for those two units comes through that one room. It's common to both units, just like the control room is common to both units.

LONG: That sorta shoots your redundancy.

Emergency procedures inside the Browns Ferry plant were also deficient. Many employees did not know what the sound of the fire alarm meant, and few had been trained in emergency procedures, despite earlier fires at the plant.

Large numbers of plant employees went into the plant control room, adding to the chaotic situation there. Instead of the six persons normally there, one assistant shift engineer reported, "the maximum number of people in the control room at any one time I guessed to be about 50 to 75."

Indicative of the tension felt in the control room is the later comment of the Unit Two shift engineer: "The plant superintendent asked me if I had control of Unit Two and if everything was O.K. *almost continuously.*"

What is the significance of the Browns Ferry incident?

One question is: What the devil were the electricians doing using a candle to test for air leaks?

Although it is perfectly possible to design an inexpensive anemometer to test for air leaks, or even use smoke from a cigarette, these methods were rejected two years ago by the Browns Ferry plant personnel in favor of using candles. Some senior personnel at the plant thought that the urethane sheet foam used to seal the cable penetrations was fireproof. The leader of the electrical conduit division at the plant said:

> The practice of using RTV-102 and sheet foam to seal air leaks has been the practice for two or three years. We believed that the urethane would not sustain a fire. Urethane samples had been tested several years ago and it needed a flame for 20 minutes to sustain a fire.

They had tested only two of the polyurethane samples, however, using an American Society for Testing Materials (ASTM) test that the Marshall Space Flight Center later found to be of marginal value. No test had been made of the foam polyurethane, however, and the NRC's consultants from the Marshall Space Flight Center found that "a cursory match test on a piece of the foam rubber disclosed almost instantaneous ignition, very rapid buring, and release of molten flaming drippings."

Even though some people at the plant thought the ASTM tests showed the penetration sealant material to be nonflammable, senior management knew it was highly flammable. The plant instrument engineer told NRC inspectors:

> During the test and startup period of Unit One [in 1973], I demonstrated the flammability of the sealing material to the Plant Superintendent. I burned the material in the Plant Superintendent's office. He immediately called someone with Construction and they discussed the situation. . . . I feel the Plant Superintendent did all that was immediately possible to investigate the situation as it appeared that Construction was not going to change the material.

The plant superintendent admitted to the NRC inspectors, "I was aware that polyurethane was flammable, but it never occurred to me that these penetrations were being tested using candles."

Many senior management personnel at the plant denied knowing of the practice of using candles to test cable penetrations.

The rest indicated that they knew candles were being used but thought the sealant materials were not flammable.

The electricians seemed to be the only group who knew both that the foam rubber was flammable and that candles were being used as the testing method. As one electrician later recounted:

> The electrical engineer called the group [of electricians] together and warned us how hazardous this method was. "Why just the other day," the electrical engineer said (in effect), "I caught some of that foam on fire and put it out with my bare hands, burning them in the process."

One of the electricians who started the fire said that candles had been used for more than two years but added, "I thought that everybody knew that the material we were using to seal our leaks in penetrations would burn. . . . I never did like it."

The real irony of the Browns Ferry fire was that two days before, a similar fire had started but had been put out successfully. After the first fire, the shift engineers and three assistant shift engineers met. According to one of them, "We discussed among the group the procedure of using lighted candles to check for air leaks. Our conclusion was that the procedure should be stopped."

Yet nothing was done. The fire was noted in the plant log and briefly discussed the next day at the plant management meeting. No one on the management level seemed to consider it a safety problem worth following up. This was the standard operating procedure; as the NRC investigative report notes,

> Previous fires in the polyurethane foam materials had not always been reported to the appropriate levels of management, and, on the occasions when reported, no action was taken to prevent recurrence.

In the face of these practices, it was probably not a question of *whether* the Browns Ferry plant would have a major fire, but *when*.

What will the fire mean for other nuclear plants? That depends on whether the NRC carries out the recommendation made by the Factory Mutual Engineering Association of Norwood, Massachusetts, the fire underwriters the NRC engaged as consultants:

> Conclusions and Recommendations:
> The original plant design did not adequately evaluate the fire hazards of grouped electrical cables in trays, grouped cable trays and materials of contruction [wall sealants] in accordance with recognized industrial "highly protected risk" criteria. . . .
> It is obvious that vital electrical circuitry controlling critical safe shutdown functions and control of more than one production unit were located in an area where normal and redundant controls were susceptible to a single localized accident. . . . A reevaluation should be made of the arrangement of important electrical circuitry and control systems, to establish that safe shutdown controls in the normal and redundant systems are routed in separated and adequately protected areas.

Every nuclear plant in the country uses a cable spreader room below its control room. Despite requirements for separation and redundancy of reactor protection and control systems, every reactor has been permitted to go into

operation with this sort of configuration, which lends itself
to a single failure's wiping out all redundant systems.

If every plant currently operating and under construction were required to rewire so as to achieve true redundancy and eliminate cable trays bunched together, I have
made calculations that indicate the cost will range between $7,680,000,000 and $12,343,000,000. It will be interesting to see whether the new commissioners of the
Nuclear Regulatory Commission will require such changes.

Except for one news release, written March 27, 1975,
NRC headquarters in Washington, D.C., has remained si-
lent about Browns Ferry. That news release, quoted below,
does not make one optimistic that any meaningful lesson
has been learned from the Browns Ferry incident.

> The functioning of some in-plant operating and safety
> systems, including emergency core cooling systems, was
> impaired due to damage to the cables.
>
> The two reactors were safety shut down and cooled
> during the fire. NRC inspectors report that there was
> redundant cooling equipment available during the
> reactor cooldown. . . .
>
> Although some instrumentation was lost, certain criti-
> cal instrumentation such as reactor water level, temper-
> ature and pressure indicators continued to function and
> both plants were safely shut down.
>
> On Unit One, although a loss-of-coolant accident had
> not occurred, the emergency core cooling system was
> activated and supplied additional water to the reactor.
> It was manually shut down to prevent overfilling. Later,
> during cooldown, when ECCS was called for manually
> as one of the several alternative means of supplying
> cooling water, it did not activate; the alternative meth-
> ods had more than sufficient capability to cool the core.

Whether the NRC has sufficient capability to cool the public's reaction—once the facts about Browns Ferry are known—will be interesting to observe.

Editor's note: Since David Comey wrote this article in 1975, no new regulations have been issued by the NRC to require changes in the way cable spreader rooms are constructed. The nuclear plants now operating in the United States were built under whatever improvised procedures for electrical cable separation the individual plant designers chose to adopt. A voluntary guideline on this subject was established by the NRC's predecessor, the Atomic Energy Commission, in 1974. Nuclear plants now under construction are generally following this guideline, but plants built before this time are exempt. However, in addition to the fact that the standard is not mandatory, it has been attacked as inadequate by numerous safety experts. The Union of Concerned Scientists, a Cambridge, Massachusetts–based organization of scientists and engineers, which evaluates advanced technologies, at this writing has a petition before the NRC which asks that the agency write new guidelines and close down currently operating plants until safety modifications can be made.

David Comey was killed in an automobile accident on January 5, 1979.

Deadly Cargoes on Main St., U.S.A.

William T. Reynolds

Much of the attention of the antinuclear movement has
been focused on nuclear production facilities themselves.
Yet every day miniature nuclear facilities, in the form of
hazardous radioactive cargoes, crisscross the roads and rails
of the United States, virtually unnoticed by the public. A
number of accidents over the past few years have highlighted the problems posed by these shipments and stimulated a move on the part of states and municipalities to
regulate this aspect of the nuclear industry.

In the early morning hours of March 31, 1977, a train
carrying four cylinders of radioactive uranium hexafluoride (a gaseous form of uranium) derailed near Rocking-
ham, North Carolina, scattering the cylinders among the
flaming wreckage. The local fire department arrived at the
site within 30 minutes. As other agencies responded, the
question arose as to whether leakage of radiation had occurred. A state of confusion existed as emergency workers
were called on and off their duties when various agencies
took lead responsibility for coordinating the response.

It was three and a half hours before the first cylinder of
uranium hexafluoride was located, and four and a half

hours before all were accounted for. A full seven hours after the wreck occurred, a federal radiological assistance team from Oak Ridge, Tennessee, finally arrived to determine conclusively that no radiological hazard existed. Another federal team from the Department of Energy's (DOE) Savannah River plant, only 165 miles away, should have arrived well in advance of the team from Oak Ridge, which is twice as far away. Because of car trouble, the Savannah River team did not arrive until eight hours after the wreck occurred.

According to the National Transportation Safety Board, at least 17 federal, state, local, and private agencies responded to this emergency. No one agency had lead authority over operations, and confusion and delay resulted. Conflicting information and delays were encountered in determining whether a radiological hazard actually existed.

Such confusion and delays are common at the site of nuclear transportation accidents. Following a truck accident involving the spillage of radioactive uranium yellowcake (a powdery processed uranium ore, see page 52) in September 1977, three days passed before adequate cleanup operations were initiated. Interstate 24 near Monteagle, Tennessee, was closed for 15 hours following the wreck of a truck carrying low-level wastes in January 1979; considerable delay was encountered in getting professional radiologists to this rural location in the middle of the night. Monteagle officials complained that the only emergency vehicle readily available is a fire truck, which is not permitted to leave the city limits. Interstate 24 near Monteagle is a major artery for the transport of radioactive wastes from nuclear power plants in the Midwest to the South Carolina burial site.

There is no federal requirement for specific emergency response plans to be developed and tested for radiological transportation emergencies, even though such plans are required for nuclear power plants. This often leaves emergency response personnel ill equipped and ill trained to respond adequately to such accidents, as evidenced by the findings of a Birmingham, Alabama, emergency response specialist, who found only two out of ten fire departments in that city with proper equipment to respond to a radiological emergency. Of the two with equipment, only one had personnel trained to use the equipment.

The International Association of Fire Chiefs supports the concept of requiring advance notification to local fire departments of hazardous radioactive shipments. Fire departments are typically the first to arrive at the site of transportation emergency situations and need to have information about what cargoes are being carried, what routes are being followed, and when the shipments are being made, in order to prepare themselves adequately for the emergency situations that do arise.

Radioactive wastes from the crippled Three Mile Island nuclear power reactor in Pennsylvania were shipped to the corners of the country in the months following the accident of March 28, 1979. In April, South Carolina health officials refused to allow the truckloads of radioactive sludge into the state after determining that the wastes were not suitable for burial at the Barnwell, South Carolina, nuclear-waste disposal site. After returning to Pennsylvania, the trucks were then directed to another nuclear waste burial site in eastern Washington.

By the end of April 1979, at least three flat-bed truckloads of Three Mile Island nuclear wastes had traveled to the Washington site through the heartland of America along Interstate 80. And more were expected.

State and local government officials, as well as private individuals, were outraged about these shipments passing through their communities—particularly so because no one had bothered to notify the police, fire departments, health officials, or in fact any public agencies along I-80 that the shipments were being made. The very agencies responsible for protecting the public's health and welfare knew no more about the whereabouts of these shipments than what was reported in the newspapers.

Although the wastes from Three Mile Island created a stir, hazardous radioactive materials from the production of nuclear power and nuclear weapons are transported every day without advance notification to public officials, without routing restrictions, and without public monitoring and inspection.

The shipment of hazardous radioactive cargoes presents a direct threat to the public. Accidents can and do happen, exposing both the general public as well as transportation workers and emergency response personnel to radiation hazards. From 1971 through August 1978, the U.S. Department of Transportation (DOT) received 369 reports of "incidents" involving radioactive materials in transit. According to DOT, about one-third of these incidents involved the spillage of radioactive materials. Although DOT requires incident reports for any mishandling problems or accidents involving radioactive materials, it has no estimate as to how many such accidents go unreported each year.

In addition to accidental releases, low levels of radiation are emitted from every package of radioactive cargo carried. Although it would be possible to build a package that would allow no radioactive emissions, the cost and weight of such a container are prohibitive. Since there is no government monitoring of these packages, the shippers of the nuclear cargoes work under an honor system to ensure that the contents are properly packaged and potentially dangerous levels of radiation are not allowed to escape.

One survey of radioactive shipments by the Los Alamos Scientific Labs, a DOE-funded research facility in New Mexico, found 1141 "occurrences" out of 2593 packages, 526 of which were of a significant nature. "Occurrences" included such problems as no labels on packages, broken security seals, improper security seals, levels of radiation higher than indicated on the package, and wrong labels posted on the packages.

Workers who handle packages of radioactive materials during transportation are the main victims of low-level

Workers load uranium fuel for shipment. Credit: AEC Photo by Frank Hoffman.

releases due to faulty packaging. Nuclear Regulatory Commission regulations do not require monitoring of truck drivers and freight handlers for radiation exposure, nor do their employers have to keep their medical records on file. An eight-state survey by the Los Alamos Labs found six out of the eight states reporting workers who received radiation doses in excess of the allowable limits, some by as much as 500 percent.

A major transportation accident would also be costly. A report to the federal Department of Energy estimated that a large-scale accident involving the release of spent fuel or plutonium in a major urban area could result in as much as $2 to $3 billion in damages due to decontamination and cleanup, relocation of residents, and loss of income. Although the probability of such an accident is low, the consequences are extreme. This estimate does not take into account the sociological, political, and economic costs that would also be associated with such a major accident. Following the Three Mile Island accident, news accounts documented significant devaluation of property in that area, a general downturn in the business climate, and negative psychological impacts.

Land held as security or on speculation would be valueless if contaminated by a major nuclear transportation accident. The Price-Anderson Act limits the liability of such an accident to $560 million. With the potential for damages well exceeding this limit, victims of such an accident might never receive full payment for their losses.

Negative economic impacts would not be limited to accident situations alone. In 1977 a Texas landowner was awarded $105,000 in compensation for property devaluation. An electric utility had built a rail spur adjacent to the landowner's property for the purpose of transporting radioactive wastes from a nuclear facility. The Texas Supreme Court upheld the landowner's contention that the possibility of an accident did exist. This would present an added risk to potential buyers of the property, resulting in a devaluation of about $300 per acre.

Channels through which local and state governments can effectively control hazardous transports do exist. Many states and municipalities across the country have become aware of the problems of transporting large quantities of radioactive materials and have decided to take action. Gaps in federal regulations provide the open door for local and state involvement.

A regulation restricting shipments of large quantities of radioactive materials enacted by the City of New York remains in effect because there is no such federal regulation. Upheld in April 1978 by DOT, this regulation paved the way for the legal involvement of other municipalities in similar regulations. More than 60 cities and states have taken the initiative to regulate and restrict the shipment of nuclear cargoes through their jurisdictions.

As a result of this movement by cities and states to take responsibility for regulating radioactive shipments, DOT has been forced to reassess its existing regulations. DOT is currently considering rule changes for the highway transport of radioactive materials. Rules established by DOT would take precedence over any local or state restrictions. However, it would be possible for DOT to incorporate provisions that would involve state and local agencies as well as the general public in planning for the shipment of radioactive materials. DOT could also allow for local and state regulation of certain aspects of radioactive shipments, such as routing restrictions, time of travel, advance notification to public agencies, and monitoring of nuclear cargoes. In addition, DOT should be petitioned to require public hearings prior to route approval for nuclear shipments, to require that emergency response plans be developed and tested, and to require the corporations making the nuclear shipments to bear some of the financial burden for providing emergency response capability in local areas.

Strong public involvement is needed to show the Department of Transportation that lax standards will not be acceptable on this issue.

Many different types of radioactive material, posing varying levels of hazard, are shipped for a number of purposes. Radioactive materials are shipped primarily by private common carriers—that is, private firms—which include passenger airlines, cargo airlines, freight trains, trucks, and ships. About 50 percent of the commercial nuclear shipments are made by trucks. Passenger airlines account for about 30 percent of nuclear shipments, which consist mainly of short-lived isotopes used for medical purposes. Cargo aircraft make up another 13 percent. Railroads are used for shipping natural uranium and uranium hexafluoride (use of railroads will increase as more shipments of spent fuel are made). Transport by ship and barge is the smallest category: ships are used mainly for transporting enriched uranium overseas. Rail and water shipments together account for about 7 percent of all commercial packages shipped. Although the least used mode, barges may offer a safer means of transporting spent fuel and high-level wastes. However, because of the added costs of building terminal facilities, few utilities are planning to utilize barges for any large quantities of materials.

The following are the major types of radioactive cargoes:

Medically related shipments.

These are primarily radiopharmaceuticals, the radioactivity from which tends to be short-lived. They are usually carried in small quantities and pose only a small level of hazard if accidentally released. A small number of shipments each year are large-quantity cobalt and cesium sources used for the treatment of cancers and other malignancies. Because of the weights involved, most of these packages are shipped by truck or rail.

About one million packages of radiopharmaceuticals were shipped in 1975. Most were carried aboard aircraft, and then by delivery trucks to hospitals. Because the level of public hazard presented by these shipments is low, most state and local governments have not established more stringent regulations than those set by DOT.

Industrial shipments.

Radioactive materials are used in certain industrial processes, principally for examining construction material, measuring the thickness of materials, and studying the geology of wells. This often involves the transport of the radioactive test devices to remote field locations. These industrial shipments involve the handling and carriage of significant quantities of radioactive material of up to hundreds of curies at a time. (A curie is the radiation emitted by one gram of radium.) Although an exact accounting is not available, about 70,000 to 90,000 such packages are shipped each year in the United States.

Nuclear-power industry shipments.

Shipments related to the production of nuclear power can be divided into two categories: "front-end" and "back-end." Front-end shipments involve the transport of various forms of uranium from mining and milling operations, through the enrichment and fuel-fabrication process, and up to insertion into the nuclear power reactor. This includes shipments of yellowcake from uranium mills, uranium hexafluoride for enrichment, and fresh fuel rods being shipped to nuclear reactors. The main radiological hazard from the shipment of these materials is from inhalation. The radiation emitted from this uranium cannot penetrate through the skin and therefore does not present an

**TRANSPORTATION ASSOCIATED WITH
NUCLEAR FUEL CYCLE**

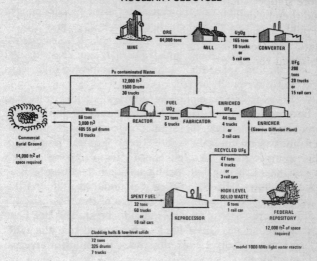

external hazard, but both yellowcake and uranium hex-afluoride can be easily dispersed and ingested if accidentally released from their packaging. Both would also present problems from heavy metal poisoning as well as chemical hazards. Fresh fuel rods, encased in metal sheathing, could not be easily dispersed and do not pose a public health hazard in the event of an accident.

Although shipments in the front end of the cycle pose only a low-level radiological hazard, shipments in the back end of the cycle, particularly the shipment of spent fuel rods and highly radioactive wastes, present a large potential for disaster. Spent fuel rods from a nuclear reactor emit lethally high levels of radiation; shipments of spent fuel rods require very heavily constructed casks to prevent the escape of radiation and to contain contents in the event of an accident.

In a dramatic series of full-scale tests, the Department of Energy has attempted to demonstrate the capability of these casks to withstand major accidents. These tests involved crashing trucks and trains carrying casks loaded with dummy spent fuel elements into massive solid walls at highway speeds and above.

However, the type of casks tested are no longer used to transport spent fuel. The casks used to ship spent fuel today are of a significantly different design and have not been subjected to full-scale testing (some Nuclear Regulatory Commission staff have been cautious in looking at such tests, pointing out that real-life accident situations can be very different from those set up under the demonstrations). The NRC licenses the designs of these casks but is not required to inspect each individual cask. The U.S. General Accounting Office has recommended that the NRC exert more control over the casks, including individual inspections.

Since there is no existing facility to dispose permanently of high-level radioactive wastes, spent fuel rods are being temporarily stored at nuclear reactor sites. As this storage capacity is being filled, pressure is mounting to move the spent fuel to make room for the storage of new wastes. In 1975, only 273 shipments of spent fuel were carried in the United States, but if our present trends continue, over 2100 shipments will be made in 1985 alone.

Lacking a permanent means of disposing of spent fuel, the nuclear industry and the federal government are developing interim means of handling it. One plan is to transfer it from one facility reaching maximum storage capacity to another having extra room. Another plan would ship spent fuel to a central facility for temporary storage. Because both plans involve increased numbers of shipments as the lethally radioactive spent fuel is moved from one temporary storage facility to another, there are more opportunities for accidents.

Nuclear-weapons production.

An elite paramilitary force of heavily armed guards traveling in unmarked and booby-trapped tractor-trailor trucks regularly carries large quantities of plutonium and other radioactive cargoes along major highways and through large metropolitan areas of the United States.

Although these shipments have the potential for causing environmental disaster in the event of a large-scale accident, they have never been subject to an open environmental impact review process. Local authorities are not informed of impending shipments or of their contents. In fact, most public officials are unaware that the shipments are even being made and would not be able to find out when they are made and what they carry even if they tried.

This special force moves above the authority of local and state legislation or police powers. If they are threatened, the armed guards have full authority to use whatever force is necessary to prevent the trailers from being occupied or waylaid.

These teams are maintained by the Department of Energy and are used for transporting materials used in the production of nuclear weapons, as well as for carrying completed nuclear warheads to deployment sites. Security is necessary to prevent hijacking or diversion by terrorist groups. At the same time, the public has a right to know what hazards are being presented by transporting such materials through their communities.

Strict secrecy does not ensure or necessarily improve the security or safety of these shipments. The Stockholm International Peace Research Institute has shown that at least 125 U.S. nuclear-weapons accidents occurred between

1945 and 1976. Of these, 32 involved major damage, destruction, or loss of nuclear weapons.

Any terrorist group intent on hijacking radioactive materials to make nuclear weapons would easily be able to discover these shipments by simply monitoring the highways near production facilities. It is not clear that absolute secrecy improves the security of the shipments, but it is clear that public health and safety are compromised by the present level of secrecy. Nuclear-weapons shipments inevitably pose two requirements for security—hijack prevention and protection of the public's health and safety. Certainly *public* security would be enhanced if local governments were informed of such shipments so they could prepare for emergency situations. Public health and safety would also be better served if a full environmental impact analysis were conducted on the shipments.

Faced with reassurances from both the nuclear industry and the federal government that the shipment of radioactive materials is proven safe, why are citizen's groups and state and local governments becoming so outspoken about this issue?

Although no catastrophic accidents have occurred in transporting radioactive materials to date, the number of shipments is increasing and with it the number of accidents. Nuclear shipments have risen from about 200,000 packages per year in 1961 to 2.5 million packages per year in 1975. By 1985, this number may easily reach 5 million packages per year. Shipments related to the nuclear fuel cycle are expected to quadruple by 1985, and the number of shipments of spent fuel may well increase tenfold by that year. The number of transportation mishaps for radioactive materials has increased from an average of 26 incidents per year in 1971 through 1974, to an average of 67 per year in 1975 through 1978.

Even when an accident releases only low levels of radiation, the chance for adverse public exposures exists, including unsuspecting motorists stopping to assist at the accident site. Public health impacts would be severe from a major accident in an urban area. Dr. Leonard Solon, director of the New York City Bureau of Radiological Health, has estimated that a large-scale release of spent fuel in New York could result in as many as 1300 early deaths from radiation exposure and perhaps tens of thousands of deaths from latent cancers.

The Association of American Railroads has been critical of the capability of spent fuel casks to survive railroad accident conditions. The Chessie Systems railroad petitioned the Interstate Commerce Commission (ICC) for relief from having to carry spent fuel shipments. Other railroads have requested permission to institute special conditions of carriage, such as speed limits of 35 miles per hour, exclusive right of way for spent fuel trains, and no other cargo on board except the spent fuel. These petitions have been turned down by the ICC.

The Airline Pilots Association has placed an embargo on the shipment of radioactive materials aboard passenger aircraft, except for those needed for medical uses. Even with medical shipments, the pilot must be notified of what is being carried and where it is located in the cargo hold.

The NRC and the DOT have overlapping authority to regulate nuclear transport. The NRC has authority for setting package-design standards for large-quantity packages and for setting security requirements for special nuclear materials. The DOT has broad authority to set package-design standards for smaller-quantity packages, labeling requirements, and notification requirements in the event of an accident. The Department of Energy and the Department of Defense set their own standards for security and handling of nuclear-weapons shipments.

Nowhere in these federal regulations are there requirements for notification to local authorities of impending shipments, routing restrictions for nuclear shipments, public involvement in planning which routes are to be used, limits on the types of quantities of particularly hazardous materials being carried through urban areas, or requirements for specific emergency response plans to be developed for transportation routes as is required for nuclear production facilities.

The secrecy surrounding the shipment and handling of nuclear weapons is slowly being cracked. The threat that secret nuclear shipments pose to our communities is being exposed. Under the National Environmental Policy Act, all Department of Energy nuclear-weapons production facilities are currently undergoing the analysis of environmental impact statements. An environmental impact analysis of commercial nuclear shipments has indicated that a large release of plutonium could result in a major environmental impact, yet no environmental impact statement is currently planned for defense-related shipments. As more local governments become aware of the specific local hazards that exist, momentum is building to open this system of transport to the full public scrutiny that is needed to make informed decisions about what risks we are willing to accept.

States and municipalities have some legal authority to restrict and regulate certain aspects of the transport of commercial radioactive materials through their jurisdictions. Citizen action is needed now to demand that this authority is exercised and strong regulations are put into effect. As the DOT and other agencies follow with federal rule changes, local communities and citizens will have to sound a strong call to ensure that local involvement is not preempted by federal regulations.

Our reliance on the production of nuclear power and nuclear weapons has generated a legacy of highly radioactive wastes that will remain deadly for thousands of generations. Even if all nuclear production facilities were shut down today, wastes stockpiled around the United States would still have to be shipped to permanent storage sites that have not yet been developed. The handling and transport of these wastes demands an absolute level of perfection which is humanly unattainable. The public should be involved in the decisions that will have to be made as to how, when, and where these wastes will be shipped. Stopping further production of these highly radioactive wastes is the wisest choice of all.

The World's Most Dangerous Garbage

Lorna Salzman

If any issue has the power to shut down the nuclear industry, it is the disposal of radioactive wastes. To the general
public it poses a more insidious and intractable threat than
any other aspect of the nuclear fuel cycle. Hostility to
dumping plans continues to mount and is hitting the industry where it hurts.

In August 1978 the Brookhaven National Laboratory on
Long Island announced that unless bans on the transport
of nuclear waste through New York City and New London,
Connecticut, were lifted, it would be forced to shut down.
Unable to ship the spent fuel elements from its experimental high-flux beam reactor for reprocessing in South Carolina, the laboratory now finds its existing storage facilities
dangerously close to overflowing. Brookhaven's problems
reflect the extent to which the waste-disposal issue has
become the Achilles' heel of the nuclear power program.
Earlier in 1978, for example, the California State legislature vetoed plans for the $3 billion Sundesert nuclear
power plant after the San Diego Gas and Electric Company failed to convince them that adequate waste disposal
facilities were available, as is required by one of Califor-

**'Just Keep Driving Around——We May Come Up
With A Solution Yet'**

Engelhardt. Copyright St. Louis Post-Dispatch.

nia's nuclear safeguard laws. The enforcement of the safeguard laws has resulted in a nuclear moratorium in California—an example that has since been followed by several
other states.

The depth of public hostility over the waste issue was
revealed by a spring 1978 Harris Poll: while residents of
New York State opposed the siting of a reactor near their
homes by two to one, they opposed the storage of radioactive wastes *anywhere* in the state by an overwhelming
four-to-one margin.

In the face of such widespread opposition, the nuclear
industry's survival clearly depends on convincing the public that it cannot only contain but also isolate the wastes
indefinitely. That distinction is important, because it is easy
to *propose* methods of waste solidification, encapsulation,
and geological burial, but it is difficult, if not impossible, to
demonstrate effective long-term isolation. Failure to do so
could well bring a national moratorium on the development of nuclear power. The industry's past record of waste
management hardly inspires confidence. Almost without
exception, the storage of wastes has been marked by
clumsy handling, incompetent inspection procedures, and
shoddy containment practices. Staggeringly large amounts
of high- and low-level wastes—in addition to plutonium—
have already been leaked (sometimes intentionally) into
the soil and water, resulting in irreversible damage to both
public health and the environment.

The real goal of these storage efforts should be successful
isolation of the radioactive materials from the biosphere,
primarily from water, people, and natural-resource deposits that may be useful to future generations. But as things
stand now, there are not even any environmental, health,
geologic, seismic, or technical criteria for radioactive waste
storage and isolation. Briefly, this means that *no one even
knows all the questions that must be asked and answered*
in order to find solutions. Without these questions and the

range of answers, it is illogical, unreasonable, and possibly insane to go on making more wastes.

URANIUM MILL TAILINGS

Milling operations crush the uranium ore, separating the uranium-238 and its small uranium-235 component from the rest of the ore and leaving posterity to deal with vast quantities of finely powdered tailings that emit the same dangerous radioisotopes as uranium itself: thorium-230 and radon-226. The latter decays to gaseous radon-222, whose radon daughters are alpha emitters that cause lung cancer if inhaled. Since the thorium-230 that gives rise to the radon has a half-life of 80,000 years (and itself arises from uranium-238 with a half-life of 4.5 billion years), these tailings will continue to give our descendants doses of alpha radiation for countless generations.

From 1948 to 1968, when uranium was mined for military and commercial purposes, about 6000 miners in the United States were needlessly exposed to dangerous levels of radioactive gases in the air of uranium mines. Several hundred have since died of lung cancer, and the U.S. Public Health Service estimates that a further 1100 deaths can be expected. In Canada dust levels in uranium mines near Ontario were consistently above the industry's safety guidelines over a period of 15 years; nothing was done to curtail them.

More than 100 million tons of mill tailings are presently stored in huge piles above ground in the West, and some have leached into rivers used for drinking water. Others have simply blown away. Incredible as it now seems, there were many cases when they were simply given away. Builders used them in foundations for schools, homes, roads, and public buildings, until it was discovered in the 1970s that the inhabitants of these buildings were receiving the equivalent of up to 500 chest x-rays each year from the radon gas seeping up through the floor. Many foundations were dug up and carted away, but by then much damage had been done. The U.S. government totally ignored the health hazards of radon in its licensing procedures until Robert Pohl, a Cornell University physicist, forced it to admit that they had been underestimated by a factor of 100,000.

The solution to the tailings problem is well known: burial to a depth of 100 feet (roughly the depth of the original body of ore) so that radon gas, with its half-life of 3.8 days, can decay before reaching the surface. Since this would drastically increase the costs of commercial nuclear power, the government plans instead to turn all tailings over to the individual state legislatures. The federal government is working to enact laws absolving itself from all damage to health caused by tailings, including those already accumulated.

HIGH-LEVEL WASTES

High-level radioactive wastes include both a liquid residue from reprocessing reactor fuel and irradiated, or "spent," fuel rods periodically removed from reactors. In reprocessing, the fuel rods (which contain millions of curies of the products of fission, such as strontium-90 and cesium-137) are chopped up, and chemicals are used to leach out the plutonium and unfissioned uranium. What remains behind is a highly corrosive, toxic solution containing all the fission products, which have an approximately 30-year half-life (half of a given quantity decays to other elements in 30 years; half of the remaining quantity decays in another 30 years, and so forth) and must be isolated from people and the biosphere for up to 20 half-lives, or 600 years. In addition, the liquid wastes contain traces of heavier elements, sometimes called actinides, which include plutonium (which has a half-life of 24,400 years), americium, curium, and neptunium. These elements emit radiation less penetrating than that of the fission products but are highly carcinogenic if inhaled or ingested. Some are extremely long-lived and require isolation for up to 500,000 years.

Today, in the absence of commercial reprocessing, spent fuel rods coming out of reactors are stored in nearby pools of water. Thus the fuel rods, containing all the fission products plus unfissioned uranium plus the actinides, are, in effect, high-level radioactive wastes, protected only by metal casing and the water. The fuel rods have high beta and gamma radioactivity, which is very penetrating, and they also have high temperatures; these two factors compound the difficulties of containment and isolation.

Moreover, there could be as many as 476,000 spent fuel rods by the year 2000, but storage is available now for only about 1 percent of these. This, then, is the problem: the technical challenge of finding ways to treat wastes, materials to encapsulate them, and a site ensured against future geophysical processes that might rupture the repository or permit leaching of radioactive materials into water. In addition, there is the sociopolitical problem of ensuring continuity and stability of institutions to guard the site from future exploration, sabotage, or war.

At present, the volume of high-level wastes from commercial reactors is only a fraction of that already generated by the military, largely through the plutonium-production program at Hanford, Washington. (It is estimated that by the 1980s, the military alone will have produced nearly a half million tons of high-level wastes measured in solid form.) Yet in terms of the radiation hazard they represent, commercial wastes are by far more dangerous. Per unit of volume, fission products from the commercial nuclear power program are 100 times greater in radiotoxicity than those produced by the military. At the end of 1977, the inventory of curies [amount of radioactivity] of important

nuclides generated from military and commercial operations was about equal. By 1985 the total inventory of fission
products in high-level wastes *alone* is expected to be 100
million curies, mostly derived from the *commercial* program. One-millionth of a curie is considered the maximum
permissible body dose.

The Hanford tanks, with about 71 million gallons of neutralized high-level liquid wastes (some in salt cake or
sludge form) have a dismal history. Over the 30 years of
military activity, 450,000 gallons of high-level waste have
leaked into the soil, and in some areas into the groundwater beneath the reservation that adjoins the Columbia
River. The leaks were largely due to the tanks being corroded by acid wastes, as well as shockingly lax inspection
and monitoring techniques. The largest single leak—115,-
000 gallons—contained nearly 270,000 curies of rutheni-
um-106, 40,000 curies of cesium-137, 4 curies of plutoni-
um-239, 0.6 curies of americium-241, and 13,000 curies of
strontium-90. Tritium and ruthenium have both been detected in the water table. Strontium-90 and iodine-131
were found in the Columbia River itself, and large amounts
of plutonium, stored in outside trenches, have had to be
dug up and dispersed because the government feared a
spontaneous atomic explosion.

Yet another Hanford leak—perhaps the biggest of all—
was kept secret for more than 20 years, to be revealed in
April of 1979. The *Seattle Post-Intelligencer* reported a
massive leak of ruthenium-103 in the early 1950s contaminated a "corridor" of land stretching from Hanford to
Spokane, Washington, more than 100 miles to the northeast. Radioactive soil, crops, and cow dung showed that the
beta-emitter was being taken up into the food chain. A
former official of the Atomic Energy Commission explained the AEC's silence over the contamination by saying that "political" conditions at the time—namely the Ko-
rean War—led to the information being classified.

Plutonium-239 also percolated into the covered storage cribs outside, and recent studies indicate that the concentration of plutonium in the sediment beneath the cribs is as high as 0.5 microcuries per gram—5000 times the maximum permissible concentration in the soil. The high-level wastes in the Hanford tanks contain up to 10,000 curies of radioactivity per gallon.

The acidic wastes at West Valley, New York (home of a now defunct nuclear-fuel reprocessing plant, the only one ever to operate in the U.S.), and at Hanford, Washington, were neutralized in order to slow down corrosion of the tanks. But neutralization creates new problems. It creates a thick sludge of fission products at the bottom of the tanks. No one knows how to get the stuff out, and there has even been talk of chopping up the tanks, sludge and all, and disposing of it in one mass. As for the small amounts of partially treated military wastes at Hanford, there appear to be difficulties with converting this ultimately to a glass form. In short, there has been some ad hoc treatment of wastes to deal with emergencies such as tank corrosion and leakage, but there still is no demonstrated technology for waste treatment and solidification. Nor is there any existing container that can guarantee nonleaching and noncorrosion for at least the requisite 600 years needed for the fission products, let alone the hundreds of thousands of years needed for the heavier radioactive elements.

LOW-LEVEL WASTES

Low-level wastes can either be liquid or solid, and include clothing, filters, tools, and other material contaminated with radiation. They have been buried in six shallow burial grounds across the United States. At the Nuclear Fuel Services site at West Valley, New York, radioactive material leached into the nearby creeks that feed Lake Erie after some of the burial trenches were flooded

with water. Later, a study undertaken by the Wood's Hole
Biological Laboratory revealed that traces of curium-244
had been found in both Lake Erie and Lake Ontario. At
Hanford, low-level wastes are *deliberately* percolated into
the soil. At Maxey Flats, Kentucky, plutonium from burial
trenches was found to have migrated 2 miles in the soil
within a few years of dumping, confounding the experts
who claimed that absorption by soil particles would prevent such movement.

A study in the June 1978 issue of *Science* reports that
trace quantities of certain radionuclides (primarily cobalt-
60, but also isotopes of plutonium, thorium, and cesium)
are migrating from both solid and liquid waste disposal pits
at Oak Ridge National Laboratory in Tennessee—despite
the predominant bedrock of the burial ground being
Conasauga shale, which is supposed to have an extremely
high absorption capacity for most fission products. The co-
balt-60 had been transported into the groundwater from
the burial trenches in biologically dangerous form, increasing the possibility of certain radionuclides—notably
plutonium-239 and americium-241—entering the human
food chain.

In addition, there have been innumerable losses of low-
level wastes through transportation accidents. In the fall of
1977, for example, 40,000 pounds of yellowcake (uranium
oxide) were spilled when a truck carrying it in metal drums
collided with a herd of horses on a deserted Colorado road
and overturned, rupturing the containers and spewing the
contents a foot deep across the road.

AGED REACTORS

What does one do with a nuclear power station that has
reached the end of its 30- to 40-year useful life? What is the
best way to get rid of the 15 to 20 percent of its contents
that is highly radioactive? As yet no one knows the ultimate

technical, financial, and health costs of complete nuclear reactor decommissioning. So far, the 20 reactors that have been closed down in the Western world have been prototypes which had been in operation only for relatively short periods of time. Their radioactivity levels are only a fraction of the levels forecast for the large commercial reactors due to be closed down in the next 20 years.

Even so, when a small research reactor was recently dismantled (under water to minimize radiation exposure) the cost of the operation was about equal to the cost of its construction. For a contemporary nuclear power plant, that would be more than $1 billion. One of America's first power reactors, the 200-megawatt Dresden Unit One boiling-water reactor in Illinois, will cost $34 million to decrud —just to have its pipes cleaned. That plant, which came into use in 1959, cost $35 million to build. No estimate has been made for the cost of completely decommissioning the plant. The decrudding is an attempt to extend the plant's useful life beyond 20 years.

Simple dismantling of a commercial atomic plant may cost anywhere from $31 million to more than $100 million in 1977 dollars—between 3 and 10 percent of the $1 billion capital cost, predicts an April 1978 U.S. Congress report, *Nuclear Power Costs.* Those figures do not include the perpetual-care costs for tending to the plant rubble, which contains radioactive nickel that may remain hazardous for up to 1.5 million years.

Reactors would be dismantled in three stages: mothballing, entombment, and complete removal—a process that for a large reactor may take from 50 to 100 years. Only five prototype plants have gone beyond the mothballing stage, during which the plant is kept intact but the reactor is sealed off with anticontamination barriers. Some schemes call for 100 years of mothballing before dismantling is attempted. That means a nuclear power plant decommissioned at the end of the Civil War would only now be unsealed for the wrecking crews.

In the second and third stages of decommissioning, according to a British study, workers tackling a British gas-cooled, graphite-moderated reactor will need to remove radioactive parts consisting of 2500 tons of mild steel, 100 tons of stainless steel, and 2500 tons of graphite. The inner layer of the concrete shield around the plant will also be radioactive to a depth of 1.5 meters. All this must be dismantled, shipped, and stored.

It may well be that the U.S. government merely intends to "mothball" all reactors. If so the American landscape will some day be dotted with monuments, even entire zones, requiring perpetual surveillance.

THE DISPOSAL DILEMMA

For the ordinary citizen caught up in the waste-disposal controversy, separating myth from reality is extremely difficult. The government persists in asserting that a solution is in hand and simply needs some hard decisions and hard money to be implemented. Yet one need only refer to some of the government's own studies to realize that what exists are not *demonstrated* technologies but merely *concepts* of waste handling, containment, and burial. In fact the more research that is done into methods of waste disposal, the more scientists are realizing the extent of the gaps in their knowledge. A recent report, prepared by the Office of Science and Technology Policy, admits that "the knowledge and technological base available today is not yet sufficient to permit complete confidence in the safety of any particular repository design or the suitability of any particular site."

No commercial waste has yet been solidified in the United States, and although some fission products have been vitrified into glass blocks in France, it has since been revealed that they have already begun to leach. Not sur-

'Nuclear waste problem? I don't know about you, but I don't want my kids growing up in a world where there aren't any problems left to solve!'

Tony Auth. © 1978, The Philadelphia Inquirer. The Washington Post Writers Group.

prisingly, such "vitrification" has come under attack. "In the opinion of the materials community, glass is relatively unstable and thermodynamically undesirable—in short it 'chews up' easily," says Rustum Roy, director of the U.S. National Academy of Science's Committee on Radioactive Waste Management, whose highly critical report was published in August 1978.

As yet no satisfactory terminal geological repository has been located. Indeed, the deadline set by the government for deep-earth isolation has already been postponed until the early 1990s, and many believe that it will be further postponed until the beginning of the next century. The main U.S. disposal strategy at present is a rather pathetic plan to construct Away-From-Reactor (AFR) sites—oversized ponds in which spent fuel elements can be temporarily stored. Without these AFRs, the nation's reactors, already rapidly exhausting their own fuel ponds, will soon have to shut down.

Such fuel pools, instead of containing solidified, contained, and geologically isolated wastes, would consist of high-level fuel rods containing millions of curies of extremely hazardous fission products. These would be protected only by thin metallic cladding and would be stored above ground in pools of water. Getting these rods from all the reactors in the Northeast to the new sites will involve dozens of shipments each year on crowded highways and through towns and cities, with constant risk of accident, collision, fire, or sabotage; one accident could expose thousands, even millions, of people to lethal doses of radiation.

In fact, the AFR policy is merely a return to an earlier, discredited concept of waste disposal known as Retrievable Surface Storage (RSS), which was intended to keep spent fuel within easy reach for eventual reprocessing. In its day, the RSS method was severely criticized by many federal agencies, which feared that it could become a permanent solution. Those fears are now confirmed. With permanent burial a distant chimera, RSS, in the guise of AFRs, will be the sole means of waste disposal for the forseeable future —a series of high-level waste repositories above ground, without the multiple safety barriers of geological sites to contain the radioactivity. Not surprisingly, critics of nuclear power see the decision to opt for AFRs as an admission that a permanent solution is remote. Or is it a cynical and none too subtle move not to solve, or even ease, the waste problem but to prop up a failing industry?

In a desperate attempt to calm public fears by digging a hole in the ground and getting some waste in there fast to demonstrate a "solution," a plan for a terminal repository—the Waste Isolation Pilot Project (WIPP)—is being pushed through in New Mexico. At first, the state (whose largest employers are two U.S. Department of Energy–funded concerns, the Los Alamos weapons laboratory and the Sandia Laboratory) welcomed the project. But support was withdrawn when they discovered that WIPP would receive not only low- and high-level *military* wastes but high-level *commercial* wastes and 1000 spent fuel assemblies as well. Even formerly enthusiastic officials are now balking at the idea, and one of them has introduced legislation in Congress designed to give states the right of veto over waste repositories. Many other senators would also like to give their states guaranteed right of veto on which the government could never renege.

But while the Department of Energy claims that it will honor any state refusal to accept waste, it clearly could not tolerate such refusals from all 50 states. Understandably, Secretary of Energy James Schlesinger is not keen to give the states a statutory right of veto: "I think the matter would be best left unresolved," he told a House committee on internal affairs.

SALT DEPOSITS

Most of the government's efforts at deep-earth burial have gone into the exploration of salt formations, primarily in New Mexico, Louisiana, New York, Ohio, and Michigan. Salt was generally viewed as the most promising of all geological media, mainly because of its plasticity, which, it was believed, could help seal the repository. As recently as 1976, officials from the Energy Research and Development Administration (ERDA—now part of the Department of Energy) were predicting confidently that burial in salt would require "only straightforward technological and engineering development." Now, however, salt is seen to have major drawbacks, all of which have been minimized by the industry: it is highly corrosive, not entirely free of water as had been assumed, and is usually located in areas of oil, gas, and potash, which could mean that there are uncharted drilling holes that would weaken the integrity of the salt formation.

This is precisely what happened at Lyons, Kansas, where

the Oak Ridge National Laboratory was storing spent fuel
in the mid-1960s. In 1970 the government announced that
Lyons would be the first federal waste repository. But over
the next few years old oil and gas holes were discovered
near the site, and the plan was abandoned. Again it was a
team of state officials, in this case the Kansas Geologic Survey, who turned the waste away. They discovered that,
contrary to the federal government's assurances, the site
had the consistency of "swiss cheese"—hardly suitable to
isolate radioactive waste for millennia.

The use of salt deposits has come in for strong criticism
from both the Office of Science and Technology Policy
(OSTP) and the U.S. Geological Survey (USGS). The OSTP
reported that while salt is the best understood of all geological media and "with conservative engineering" might be
an acceptable repository, it has unique problems: "Because
of salt's highly corrosive nature, currently planned waste
containers would seem to be breached and substantially
corroded by all but the very driest salt within months to
years." They add that salt is soluble and "does not provide
the absorptive qualities of other rocks nor is it benign to
interactions with the waste and container." These factors,
it states, could prove "troublesome" in the event of a canister breaking. The OSTP also stresses the great gaps in
technical knowledge of waste disposal.

The same point is taken up in a USGS circular on geological disposal: "Many of the interactions (between waste,
canister, and geological medium) are not well understood,
and this lack of understanding contributes considerable
uncertainty to evaluations of the risks of geological disposal
of high-level waste." The circular also pinpoints three
major problems that are likely to occur in salt formations:
disturbance of the medium caused by the actual mining;
chemical disturbances created by introducing new fluids
not in chemical equilibrium with the salt; and thermal
disturbances from hot wastes that will in turn compound

the two other problems. It also expressed concern for unknown geological faults, groundwater conduits, and abandoned excavations—all of which could allow water into the repository. In addition, hot canisters tend to attract corrosive brine toward them.

Salt was not the only geological medium the USGS was worried about. In rock deposits, chemical changes due to the introduced heat could lead to thermal expansion and contraction that would fracture the canisters. This thermal energy could also break down hydrated minerals and form new ones, with significant increases in the permeability of the rock. "Given the current state of our knowledge," warns the USGS, "the uncertainties associated with hot wastes that interact chemically and mechanically with the rock and fluid system appear very high."

In June 1978 a brutally honest report from the U.S. Environmental Protection Agency gave what may be the death blow to the use of salt for disposal. This report explodes the common belief that many salts do not contain water. Close inspection of even the driest salts reveals "significant amounts of water in fluid inclusions and intergranular boundaries." The waste canisters are "likely to be bathed in water soon after emplacement" and, worse still, the moisture will actually cause the crystals to burst at temperatures half that of the canister. As for the canister itself, the report states that "no tests . . . have shown that any of the candidate metals will resist corrosion by the salt solutions that are likely to be at the canister surface for a significantly long time. Under these circumstances it is likely that the canister could be breached *within time scales of a decade or less.*" The mounting evidence of salt's unsuitability for long-term waste disposal prompted the influential Interagency Review Group on Nuclear Waste Management in March 1979 to urge exploration of different geologic media in the search for a waste solution.

CONCLUSION

Neither the government nor the nuclear industry will countenance discussion of a nuclear moratorium until waste-isolation technology has been demonstrated, nor will they admit that continued production of wastes could conceivably make the situation worse. They respond that even if the industry shuts down, we will still have large amounts of waste to deal with. This is an argument that totally misses the point: not only is it easier to deal with a fixed quantity of wastes than with a quantity 10 times as large, but also there may be very few—perhaps only one —geologically acceptable burial site in the United States. Only a limited amount of waste could then be accommodated, and continued production will require additional burial sites that may be totally unsatisfactory.

The key questions are: How much is the problem compounded by *not* stopping waste production? How many

Richard Willson. By permission of Friends of the Earth.

"Presumably a shrine for one of their primitive religious cults."

tons of uranium tailings will blow in the wind? How many more thousands of annual truck and rail shipments of uranium and spent fuel will be needed? How many more derailment accidents will there be? How many additional AFRs must be built? And how many permanent burial sites?

If after 20 years of nuclear power no single example of effective containment has been demonstrated, what hope is there of future success? Can there possibly be any justification for allowing the nuclear industry to go on manufacturing waste products whose potential for destruction neither scientists nor government can begin to calculate? Can it be permitted to prop itself up with the myth rather than the reality of safe waste disposal?

A Little Radiation Goes a Long Way

Kitty Tucker

The nuclear power and nuclear medicine industries were founded on the assumption that a little radiation won't hurt anybody, but studies conducted since the "atomic age" was introduced at Hiroshima in 1945 have shown that radiation can kill in ways other than atomic explosions. Low levels of ionizing radiation are already causing increased incidences of cancer in persons exposed to atomic-weapons testing, radiation on the job, and radiation from medical x-rays.

The energy of different types of radiation is generally divided into two categories: ionizing and nonionizing. Ionizing radiation is capable of producing ionization, or actual structural changes, within an atom. These changes can damage living cells.

According to Dr. Karl Morgan, a founder of the health physics profession in America, "there is no safe level of exposure and there is no dose of radiation so low that the risk of a malignancy is zero."

The radioactive products of the nuclear power program make its characterization as peaceful inappropriate. The ranks of the antinuclear movement have swelled as more people understand the long-term implications for increased cancer, diseases of aging, and genetic damage to future generations. Workers in the nuclear industry already suffer from these problems.

PAST HISTORY

The cancer danger to workers exposed to ionizing radiation was discovered in the 1920s when radium-dial painters began to die of cancer. In the early 1900s, young women were hired to paint watch dials with radium paint. It was customary for a worker to place the tip of the paintbrush in her mouth to get a point for producing the tiny numbers. The daily ingestion of radiation eventually led to cancer.

Radiologists were also early victims of radiation. Prior to 1950, when awareness of radiation risks was low, excess mortality among radiologists ranged from 60 percent for heart disease to 600 percent for leukemia.

Early standards were set on the assumption that ionizing radiation did not harm if it was below a certain exposure level, but the level of exposure regarded as "safe" has been dropping. Before 1950, scientists assumed that this safe level was equivalent to about 52 rems per year. (A rem is a unit that has been set up as a measure of biological damage from radiation.) Then a new exposure level of 15 rems per year was recommended by the International Commission for Radiation Protection. Reduction was recommended again in 1956 to the current level of 5 rems per year for workers. In 1975 and 1978, respectively, the Natural Resources Defense Council and cancer researcher Dr. Rosalie Bertell of the Roswell Park Memorial Institute in Buffalo, New York, petitioned the Nuclear Regulatory

Commission (NRC) to reduce this exposure level tenfold to ½ rem.

Two million workers, hundreds of thousands of military personnel, and large numbers of the general public needlessly exposed to x-rays are serving as involuntary guinea pigs for the study of radiation effects on people. Since radiation cannot be seen, heard, smelled, or touched, people are not even aware of its presence. Radiation-induced cancer has a latency period of from five to 30 years between the time of exposure and the time the disease appears, so the association between exposure and injury is difficult to prove.

RECENT STUDIES, COVERUPS, AND ATTEMPTED COVERUPS

Dr. Alice Stewart, a member of Britain's Royal College of Physicians and a founder of the study of radiation epidemiology, was one of the first scientists to revolutionize thinking about low-level radiation. In 1958 she reported that of 1800 British children who died of cancer before age 10, twice as many died whose mothers had been given abdominal x-rays during pregnancy. After extensive controversy, her findings were recognized and obstetrical practices began to change.

Dr. Irwin Bross of the Roswell Park Memorial Institute found that increased childhood leukemia and significant genetic degradation were associated with low-level ionizing radiation in a research effort known as the Tri-State Study. The Tri-State Study followed a population of over 13 million people from a three state area who received recorded medical and dental x-rays. Dr. Bross's research showed that the National Cancer Institute's (NCI) Breast Cancer Detection Demonstration Program, which involved extensive x-rays of women, would produce more cancers than the NCI could possibly cure.

Funding from the National Cancer Institute was terminated after nine years when Dr. Bross became vocal in professional journals and before Congress in criticizing unnecessary radiation exposure by doctors.

Dr. Thomas Mancuso of the University of Pittsburgh studied the death certificates of former workers at the Hanford nuclear facility in Washington state. (Hanford is the principal producer of plutonium for the nuclear-weapons program.) The study, involving a data base of over 35,000 workers, showed increases in several forms of cancer among the Hanford workers, which occurred at exposure levels 10 to 20 times below federal standards.

Dr. Mancuso, working with Dr. Alice Stewart and statistician George Kneale, first presented his findings in 1976 after the Department of Energy (DOE) terminated his research funding. Mancuso refused to publish his data prematurely in 1974 when the government tried to refute an earlier study conducted by Dr. Samuel Milham, which showed a 25 percent cancer excess at Hanford. Dr. Sidney Marks, then working for one of DOE's predecessors, the Atomic Energy Commission (AEC), initiated a two-year termination plan for the Mancuso study. Dr. Marks then left the government and now runs the research project formerly headed by Mancuso.

The nuclear establishment, alarmed by Mancuso's independence, initiated an international whispering campaign in an effort to discredit Dr. Mancuso by anonymous criticisms attacking him and his work. The researchers were never confronted directly with these criticisms.

Bob Alvarez of the Environmental Policy Center was alarmed by the funding terminations of the Mancuso and Bross studies, and he set out to get their funding reinstated. In the fall of 1977, he sent letters signed by Anthony Mazzocchi, vice president of the Oil, Chemical and Atomic Workers Union, and seven national environmental organizations to the appropriate federal agencies demanding an explanation for the funding terminations. He also began lobbying for a congressional inquiry.

Congressman Paul Rogers, a Democrat from Florida, chaired hearings of the Commerce Subcommittee on Health and the Environment in early 1978 that explored Mancuso's funding termination. Officials offered various stories for the termination.

A claim by DOE officials that Mancuso was retiring proved false; he had several years to go before reaching the 70-year-old retirement age at the University of Pittsburgh. DOE officials also claimed that Dr. Mancuso received bad peer reviews, but when the officials were forced to produce the reviews, only Dr. Sidney Marks's review was negative. The other four disinterested reviewers recommended continued funding on an expanded basis.

Dr. Bross told the Rogers hearings that his research showed that the dose at which cancer incidence doubled was inside the 5-rem range. Instead of changing standards and practices according to the new scientific information, the government tried to deny it.

"In summary," Dr. Bross testified, "the 'Big Science' federal agencies such as the AEC, their industrial constituencies, and their allies in the engineering, scientific, and medical communities have been lying to the public about the hazards of low-level ionizing radiation for about 25 years. To protect the lie from exposure by honest researchers, 'Big Science' used its control of the 'peer review' machinery to suppress, vilify, or cut off the funding of the 'little scientists' who told the truth."

The Rogers hearings also explored cancer and leukemia findings among former military and civilian personnel exposed to atmospheric atomic-weapons testing in the 1950s. Dorothy Jones of Another Mother for Peace testified that their effort to locate victims of the Department of Defense testing programs had uncovered numerous cancer victims. She said that these victims, former members of the armed

'Not to worry . . . just a little fallout from our 'fifties weapons tests.'

Tony Auth. © 1979, The Philadelphia Inquirer. The Washington Post Writers Group.

forces, were being denied any compensation.

Congress passed special legislation to compensate the Marshall Islanders affected by the fallout from U.S. H-bomb tests in the Pacific. (The Marshall Islanders experienced increased incidences of thyroid tumors and related problems.) Why, Ms. Jones asked, did the government steadfastly refuse to recognize and pay service-connected claims submitted by our own men who participated in these same tests?

The hearings revealed that troops were exposed to fallout intentionally. Scientists attended some tests in full protective gear while troops faced tests wearing as little as shorts. Radiation film badges worn by troops were found in storage; they had never been sent to a laboratory for readings of the soldiers' exposure.

Former serviceman Artie Duvall, Jr., was told that he had been exposed to lethal radiation during the tests and that if he revealed any of the facts regarding this event in the next 10 years, he would be jailed for 10 years and fined $10,000. He was later denied benefits when he contracted cancer, partly on the ground that he waited so long to apply, since he kept his oath of secrecy.

The Rogers subcommittee demanded an explanation for the fact that the army had only one major working one-fourth time on locating the ex-service personnel exposed to radiation. Eight of the 2245 soldiers who participated in one 1947 nuclear test nicknamed "Smoky" developed leukemia. Leukemia is a rare disease, and the eight cases are 50 times the average number expected for that age.

Dr. Thomas Najarian of the Boston University School of Medicine, in a nongovernmental study of the Portsmouth, New Hampshire, Navy Yard nuclear submarine workers, reported a 450 percent higher death rate from leukemia among the radiation workers than in the general population. Surveying the deaths of 146 people who had worked at the shipyard, Najarian found that 38.1 percent had died of cancer—double the national rate.

Ronald Belhumeur of Dover, New Hampshire, a civilian employee at Portsmouth for 30 years, told the Rogers subcommittee about the aftermath of a contaminated water spill from the first nuclear submarine, the *Nautilus*. "The crew that I worked with on the *Nautilus* are all dead," he reported. "I am the last one. The machinist died of cancer only months after the spill. My fellow workers, tank cleaners, died from cancer and one from natural causes." In 1977 Belhumeur's superior and a machinist supervisor died only months apart from leukemia. They had investigated the cause of the spill, and Belhumeur said they must have been "belted" with radiation.

OTHER DANGERS

Cancer is not the only problem resulting from radiation exposure. Dr. Bertell has discovered that radiation exposure seems to speed the aging process, consequently heightening susceptibility to diseases associated with aging.

Rapidly dividing cells are highly susceptible to radiation damage. Thus infants are more susceptible than adults, and the embryo and fetus are more susceptible than an infant. In 1963 the data of Dr. L. H. leVann showed that each radioactive atom is some 10 million to 100 million times more toxic to developing embryos than a molecule of the most potent deformity-causing substances such as Thalidomide. Radiation standards are targeted to the infant organs, rather than to the embryo or the fetus.

The current standards allow 5 rems per year for nuclear workers and 0.17 rem per year for the general public. Workers are allowed to receive 30 times more radiation, apparently on the assumption that workers are not entitled to a safe workplace. Obviously, they are not more resistant to radiation dangers. The annual 5-rem worker limit can be ignored under a "Catch-22" clause. Under certain conditions, an individual working in a restricted area can receive a whole body dose of up to 3 rems per calendar quarter, or 12 rems annually. *This dose is 70 times greater than that for the general public.*

These worker exposures greatly increase the risk of genetic damage to the workers' children. Drs. John Gofman and Arthur Tamplin assembled a vast body of data indicating that if the radiation doses for workers were to be received by the entire U.S. population as a result of peacetime use of nuclear energy, there would be at least 32,000 and perhaps as many as 64,000 additional deaths each year! These figures did not include fetal and infant mortality, genetic defects, or any more subtle long-range effects on health.

The risks to the unborn fetus are 200 times greater than the risks to a 50-year-old man, according to the National

Academy of Sciences. The NRC has refused to take any action to protect the unborn beyond issuing a special Regulatory Guide warning women workers to take certain precautions if they become pregnant. According to the NRC, the woman should ask to be reassigned, leave her job, delay having children, or take the increased risk. The greatest fetal risk is during the first three months of pregnancy. The mother may not even be aware she is pregnant until it is too late.

Nuclear workers are not the only people facing radiation dangers. The routine use of diagnostic x-rays for medical and dental purposes always involves some risk to the patient. Although doctors and dentists are expected to help patients evaluate the risk of the x-ray against the diagnostic benefits, many of them are too uninformed to provide such guidance. Only one state, California, even requires questions relating to the effects of radiation on people on the state board examination for doctors. Only four states, New York, New Jersey, California, and Kentucky, require education, training, and certification of x-ray technologists. There are no safeguards to keep exposures to patients as low as possible, and exposure information is often not recorded on the patient's chart. Many physicians don't even know the dosage received from x-ray equipment.

Dr. Irwin Bross estimates that between one-third and one-half of all x-rays currently given are unnecessary or useless, and this is now confirmed by official estimates. At least two out of three x-rays use two to ten times the minimum dosage that is feasible with up-to-date equipment. At least 5 percent of the U.S. population receives an adverse effect from x-rays.

"The public will have to learn to protect itself," says Dr. Bross, "because at present there is no way to stop these exposures to unnecessary and excessive radiation other than for the patient to refuse it."

Uranium miners. Credit: Department of Energy.

HAZARDS OF THE NUCLEAR-FUEL CYCLE:
MINING

The first step toward nuclear power or nuclear weapons is the mining of the uranium essential for the nuclear reaction. Most of the uranium in this country is found in the Colorado Plateau in the Southwest. The danger to miners lies in the airborne radon gas present in the uranium mines. Radon decays, producing a series of isotopes called "radon daughters." These radioactive particles emit alpha radiation. When inhaled, they lodge in the miners' lungs, leading to small cell carcinoma for many of them. European nations first found that the cancer danger to miners could be reduced by ventilating the mines for the price of about 1 percent of the total cost of the mines.

The Atomic Energy Commission was the sole purchaser of uranium in the U.S. from 1946 through the mid-1960s. The Atomic Energy Act gave the AEC control over "source material after removal from its place of deposit in nature." The AEC interpreted these words to mean after shipment from the mine facility, rather than removal from its natural deposit inside the mine, and refused to set any standards to protect workers. Consequently, uranium miners were exposed to levels equal to or higher than those present in the European mines before ventilation.

The Oklahoma-based Kerr-McGee Corporation leased the mineral rights on the Navajo Reservation outside Red Rock, Arizona, in 1949. It put unemployed Navajos to work there for 90 cents an hour, less than the minimum wage. Kerr-McGee didn't tell the miners anything about radiation and didn't install any ventilation.

"Those miners up at Cove had 100 times the levels of radioactivity allowed today," according to LuVerne Husen, director of the Public Health Service in Shiprock, Arizona. "Inside, the mines were like radium chambers, giving off unmeasured and unregulated amounts of radiation."

Husen added, "The problem was that back in the 50s, nobody was riding herd on the companies. The uranium mine operators got what they could as quick as they could out of those mines. They sent anybody who was old enough to hold a shovel and handle a wheelbarrow into the mines to cart the stuff out," Husen told Tom Barry of the *Navajo Times*.

Kerr-McGee refuses to pay compensation to the victims themselves or to the surviving families of the mining victims. Twenty-five of the 100 miners in the Red Rock chapter have already died of cancer, and more are feared to be dying. In addition to lung cancer, the miners suffer from chronic bronchosis, emphysema, pulmonary fibrosis, and other nonmalignant respiratory problems.

PROCESSING

Workers in uranium conversion, enrichment, and fabrication plants also face the danger of inhaling uranium particulates that can cause lung cancer. There are two existing commercial uranium-hexafluoride production facilities, three government-owned enrichment plants, and nine fuel fabrication plants. The *Atlanta Journal* has reported that cancer deaths in the area surrounding the Jonesboro, Tennessee, fabrication plant have doubled in the 20 years since the plant opened.

Workers in plutonium fuel fabrication plants face even greater danger, as plutonium is far more toxic than uranium. Swallowing it in a quantity that can be seen would sear the digestive tract, killing quickly and painfully. Microscopic amounts can lead to cancer years later. Plutonium stays deadly for at least a quarter million years. Once it has escaped into the environment, it cannot be recaptured or destroyed.

Workers at the Kerr-McGee plutonium fuel fabrication plant in Crescent, Oklahoma (where Karen Silkwood worked, see "Waving Goodbye to the Bill of Rights" page 162) were poorly trained in the latter years of the facility's operation. Workers have stated that the plant was well equipped when it opened, but in later years equipment broke down and was not replaced. The company policy often neglected worker health and safety to cut costs or speed production.

The Environmental Policy Institute in Washington, D.C., has initiated a long-term health followup study of workers at the Kerr-McGee uranium and plutonium fuel fabrication plants outside Crescent. The study seeks to identify any common health problems that occur among the workers and may provide a data base to assist workers in subsequent claims for compensation. The study is being expanded to include other nuclear workers who wish to participate.

WEAPONS PRODUCTION

Weapons facilities also pose dangers to workers. The Rocky Flats Nuclear Weapons Facility is located 16 miles upwind of downtown Denver, Colorado. Approximately 2750 employees are involved in manufacturing nuclear-weapons components and repairing defective bomb and warhead components. Since the facility opened in 1953, over 200 fires have occurred, some of them releasing radioactive particles in the surrounding community. Fires are a problem at the facility because plutonium ignites spontaneously when exposed to air. Routine releases have caused alarming increases in the levels of radiation in the area. As of 1974 there had been 171 plutonium-contamination cases at Rocky Flats. Contamination has even been found in the cafeteria. On September 10, 1958, surface smears (tests made to determine levels of radiation) were taken; 50 out of 54 showed contamination above tolerable limits. Ninety-seven out of 99 smears in the locker room showed contamination.

POWER PLANT OPERATION

Gertrude Dixon, research director of the League Against Nuclear Dangers in Wisconsin, reports that 122 workers at power reactors and 79 in fuel processing and fabrication received more than the allowed 5 rems per year in 1974, according to federal records. When she compared the

NRC figures for 1975, she found an increase of 158 percent in those receiving over 5 rems. These are just the exposures actually reported.

Workers at the nuclear power plants wear film badges and carry dosimeters to record exposure to radiation. Neither device detects alpha radiation, so exposure to alpha radiation must be determined from air samples.

Anthony Mazzocchi, of the Oil, Chemical, and Atomic Workers Union, points out that the radiation devices must be read and reported in order to aid the workers. He tells of an early experiment by a nuclear worker who placed his film badge next to a radiation source for a couple of days and then turned the badge in with his name. He was never contacted about the overexposure.

A 1978 NRC memo revealed that workers at some nuclear power plants are being exposed to unexpected neutron radiation. The film badges in use were not identifying the neutron exposures of up to several rems in a few hours or days. Workers in the North Anna nuclear power plant in Virginia discovered that their dosages were 18 times higher than registered when neutrons were factored in.

Maintenance and repair personnel face greater dangers at nuclear power plants, as they are more likely to be needed in high-radiation areas. Refueling at a nuclear plant is a particularly dangerous time, as the used fuel rods taken from the reactor are exceedingly "hot."

REPROCESSING

Plans to move to a breeder economy, using the more toxic plutonium as a reactor fuel, pose serious problems for the potential work force. The only U.S. experience with commercial reprocessing occurred at the Nuclear Fuel Services plant in West Valley, New York. The dosage to workers consistently increased from 1966 until 1972, when the plant closed. Releases to the outside environment also increased over the six-year period of operation.

Hazards to temporary workers at nuclear facilities may be of much greater magnitude, because they are often hired for the "dirty" jobs of repairing leaks, refueling, and cleaning up spills. These workers are hired to avoid "burning out" (overexposing) the regular, skilled workers. Between 1969 and 1972, a total of 54,675 persons worked at different nuclear facilities, according to the AEC. Of this number, 30 percent were employed for less than 90 days. The potential damage to the national gene pool from these heavy exposures to temporary workers is cause for alarm.

At the West Valley reprocessing plant, temporary workers outnumbered the permanent staff by more than ten to one. These workers were used for repair jobs in "hot" areas. According to other workers, these temporaries were told virtually nothing about the hazards of their jobs. A worker might be sent in to turn a bolt and receive the maximum dose in a few minutes. The worker would be paid for half a day but not told that he or she faced an increased risk of genetic damage or deformed children.

West Valley workers faced the same dangers of plutonium contamination as the Kerr-McGee fuel-fabrication workers. Interviews with workers revealed a similar callous management attitude toward worker safety: frequent overexposures, inadequate training, and the consequent worker inability to appreciate the hazards.

Periodically the lunchroom tables at West Valley were found to be contaminated with alpha radiation. Grossly inadequate alpha monitoring was set up at plant exits, so workers often went home contaminated. On one occasion contaminated items in a worker's home included a baby blanket, clothing, and furniture. During periods of high rainfall, contaminated holding lagoons were frequently dumped into the creeks.

WASTE STORAGE

With or without reprocessing, the handling of the final nuclear waste remains an unsolved dilemma. No sites for long-term, permanent waste disposal have been established. DOE plans for a test disposal site near Carlsbad, New Mexico, face stiff local opposition. The site did not meet original DOE criteria, so the criteria were changed.

A 1973 NRC report revealed that waste workers receive far higher doses of radiation than the average nuclear worker. In 1970, for instance, 39 percent of all radiation-disposal workers received doses above the 5-rem-per-year limit. In 1971, about 60 percent of these workers received more than the limit. Recent evidence has shown increased danger of cancer in the 1-rem range of exposure. In 1970, there were 63 percent of waste workers who received more than 1 rem, and in 1971, there were 73 percent.

Major military "temporary" waste sites are found at Hanford, Washington; Savannah River, South Carolina; Idaho Falls, Idaho; and Oak Ridge, Tennessee. Commercial-waste storage sites exist at Maxey Flats, Kentucky; Morris, Illinois; Beatty, Nevada; West Valley, New York; Barnwell, South Carolina; and Richland, Washington.

NUCLEAR MATERIALS TRANSPORTATION

In addition to the dangers posed at each nuclear facility, radioactive materials must be transported across the country. Major radioactive transportation routes crisscross the nation. Accidents are possible along every route, and some of the routes take materials through heavily populated areas and major cities.

RADIATION DANGER THREATENS EVERYONE

The Occupational Safety and Health Administration has
no authority over radiation control. Radiation regulation is
under either the NRC or the DOE. Unions would like to
see this changed, because the NRC rarely levies penalties
for overexposures to workers.

While workers at nuclear facilities face the primary dangers of radiation, residents near such facilities are exposed
to both "routine" and accidental releases of radiation.

These releases may eventually travel through air or water pathways to distant locations. "A little radiation goes a long way," notes Bob Alvarez. "If the little bit of radiation escaping from a nuclear facility causes cancer in you, or in your child, you won't be comforted by the fact that it was just a little bit."

Radiation Explained

When it comes to radiation, many people already suffer from an overdose—of information. News of millirems, whole-body doses, and plumes of radioactivity can become a mass of obscure and often contradictory data.

"Experts" cluster at every point on the radiation-controversy spectrum. It's safe. It's not safe. It may be safe. Look, if the "experts" don't know what it all means, how are we supposed to figure it out? Actually, a great deal *is* known about radiation, what it is, and how it can harm us. So much concrete evidence of the threat to life posed by radiation is known that a substantial part of the scientific community is speaking out—sometimes in esoteric language, sometimes eloquently—to put an end to radioactive contamination of the Earth.

What's known is very depressing. Radiation, even in the smallest doses, can cause incurable, painful, and often fatal cancer. It can induce horrid genetic defects that could be passed on for all the generations of human beings to come. It can act as a virtually permanent pollutant—one that can never be "cleaned up." It is being added to the environment in unknown amounts *deliberately* through ordinary operation of nuclear power plants and their attendant fuel cycle.

What follows is an attempt to outline in simple question-and-answer form, some of the basic properties of radiation and what it is doing to us.

What is radiation?

Radiation refers to particles and energy that radiate from unstable atomic nuclei. (This is discussed in detail in the article titled "Splitting Atoms to Boil Water.") The most dangerous type of radiation is called ionizing radiation. It gets its name by changing atoms it encounters into ions—atoms that have had some of their electrons ripped away, leaving a fractured atom that now bears an electric charge. This is done with the atomic equivalent of brute force. Alpha particles (which are identical to helium nuclei), beta particles (electrons), and gamma rays (high-powered x-rays) physically tear apart atoms that lie in their path. Gruesome as this sounds, in and of itself, it is painless. It's the *results* of ionization that cause suffering.

Where does radiation come from?

A brilliant nuclear-industry advertising campaign a few years ago featured a kindly old woman proudly holding her latest baked creation. "Mom's apple pie is radioactive," read the headline. "So is Mom," it concluded. Yes, there's hardly anything you can think of that is not radioactive to some degree—the book you are now holding, for example. The radiation emitted by objects and the Earth, and that which filters into the atmosphere from cosmic rays, is collectively called background radiation. This naturally occuring radiation is one of life's built-in hazards. It is, for now, impossible to avoid.

Man-made radiation is another matter. In terms of volume, the most prevalent form of this type of radiation comes from the medical industry in the form of x-rays used to diagnose or treat illnesses. Every time your dentist lays that lead blanket over you and scurries from the room, you are being dosed with radiation.

Nuclear power and weapons production is the second biggest source of man-made radiation. At every step of the nuclear fuel cycle, cancer-causing poisons escape into the air, soil, and water. By the way, radioactive fallout—the permanent calling card left behind by atmospheric nuclear explosions—is now factored in as part of the "natural background radiation."

Atomic power proponents are quite accurate when they say we swim in a sea of radiation every day. Unfortunately, the sea claims many victims every year.

How does radiation hurt us?

Two scientists who helped pioneer the field of radiation health, Drs. John W. Gofman and Arthur R. Tamplin, came up with the best description of what radiation does. They liken it to "jagged pieces of shrapnel passing through tissues" of living things. Radiation can puncture or destroy cells. It can leave trails of disruptive ions as it passes through your body. It can smash the chromosomes that carry the cell's instructions for living and govern the continuation of the species. Even worse, it can cause just a little damage in a chromosome, just enough to trigger uncontrolled cellular reproduction—a cancer. Or, just enough to damage a gene or two, the microscopic chemical strands that determine what offspring will be like, and cause a mutation. There are no known cures for the results of these "little" damages.

Assessing the extent of biological injury done by a given quantity of radiation is a hairy business. Radiation-caused biological damage is counted in rems, a measure of the amount of energy absorbed by living tissue, multiplied by a figure that represents how harmful a particular type of radiation is. The unit most commonly discussed is the millirem, a thousandth of a rem. Average background radiation at the Earth's surface is 100 millirems per year. Federal exposure standards allow average public exposure up to 170 millirems per year—not including medical exposures. Nuclear power plant workers, however, can be exposed to 5000 millirems each year and in some cases up to 12,000 millirems.

So much for numbers. In a way, the figures are irrelevant. You see, there is no safe level of ionizing radiation. There is no threshold of exposure (170 millirems, or what have you) below which there is no danger of cancer or genetic mutation. This is so, says Dr. Helen Caldicott, because, "it takes only one radioactive atom, one cell, and one gene to initiate the cancer or mutation cycle. Any exposure at all therefore constitutes a serious gamble with the mechanisms of life." Even innocuous-sounding "background" radiation is responsible for 19,000 cancer deaths and 588,-000 genetic defect-related deaths each year in the United States alone, according to Drs. Gofman and Tamplin. When the nuclear industry says it adds only radiation equal to or less than background levels (itself a dubious claim), they are adding to that death toll.

The story is told of Leo Szilard, a Hungarian scientist who fled Hitler's Europe for England in 1933. Szilard was waiting at a traffic light one day that September, and before the light turned green, he figured out how nuclear fission could start a self-sustaining chain reaction. He spent the next few years trying to keep this explosive secret quiet in a world rapidly moving toward unparalleled global war. He was unsuccessful. After Hiroshima and Nagasaki withered beneath American nuclear bombs, Szilard turned to biology, the study of life. It is at the biological level that radiation deals death.

There are dozens of radioactive substances. Each emits different amounts of the different types of ionizing radiation for different lengths of time. Some are more dangerous than others, many differ only in that they cause cancer in different parts of the body. Unstable as they may be, radioactive atoms closely resemble ordinary elements that are essential to life. When these impostors enter the body, they are swept into the conventional biological machinery, where, depending on what element they are masquerading as, they can quietly kill you.

Gamma rays—the high-energy electromagnetic waves that resemble x-rays—can originate outside your body and disrupt cells as they pass right through you. Alpha and beta particles pose their greatest threat when they are emitted inside your body. By swallowing or inhaling alpha or beta emitters, you give entry to substances that can lodge in various body organs and systems and irradiate nearby cells with intense levels of ionizing radiation. Iodine for example, is used to make thyroxin, an important hormone manufactured by the thyroid gland. Most of the iodine we consume, radioactive or not, is absorbed by the thyroid gland at the base of the neck for use in thyroxin production. Radioactive iodine, usually iodine-131, settles in the thyroid gland and emits beta radiation to nearby cells, killing some and damaging others. After a few years, some of

the damaged cells may run wild—multiplying at a dangerous rate and signaling the start of a potentially fatal thyroid
cancer. Other radioactive substances target your bones,
where they can start bone cancer or leukemia. Some radioisotopes track for the gonads, where they can irradiate
vulnerable sperm and egg cells, causing genetic mutations
that would be passed on to any offspring, assuming those
offspring survive. One such radioactive substance is
plutonium-239, an intense alpha emitter that can also
lodge in the lungs. Alpha particles, outside the body, can
be stopped by a piece of paper. Inside the body, they are
the most dangerous radioactive emission, acting like tiny
cannonballs that smash cells and cripple chromosomes. Silently.

The fact that different radioisotopes migrate to different
parts of the body gives rise to some of the confusion surrounding radiation. Industry and government officials like
to talk about "whole-body doses" of radiation—that is, av-
eraging the radiation's effect over your entire body. This
way of looking at it leads to misleadingly low estimates of
the danger posed by radioactive emission from, say, leaky
nuclear power plants. A whole body dose of iodine-131
may be relatively insignificant, but you have to remember
that most of that dose will be concentrated in one small
body organ, and there, probably in just a few precious cells.
It only takes one emission striking the proper gene to trigger a fatal cancer years after the initial exposure.

How does radiation get to us?

Every way imaginable. When you breathe, you can inhale alpha-emitting radon daughters given off by the
mountains of uranium mill tailings left exposed in the
West. When you drink milk, you can consume iodine-131,
which settles on pasture grass eaten by dairy cattle who
concentrate the radioisotope in their bodies and pass it on

in their milk. When you eat fish, you can take in huge amounts of cesium-137, a gamma emitter that resembles potassium, an important part of every living cell. Because fresh water is relatively low in potassium, fresh-water fish biologically concentrate that essential element. When cesium-137 from an upstream nuclear power plant enters the ecosystem, it too is concentrated by the fish. The concentration is so strong that volume for volume, eating those fish would expose you to 1000 times more cesium-137 than drinking the water they swim in.

Obviously, we cannot stop breathing, drinking, and eating just to prevent ourselves from being exposed to radioactive poisons that can give us cancer. The logical thing to do is to minimize the addition of radiotoxins to the environment. This is what stopping nuclear power is all about.

What kind of time frame are we working with?

A very short one. We are only one human generation into the atomic age. Cancer outbreaks in groups of people exposed to heightened levels of radiation are already well documented. Leukemia epidemics erupted in the years following the Hiroshima and Nagasaki attacks. Leukemia deaths among U.S. soldiers intentionally exposed to atomic weapon fallout are higher than normal. (Nuclear-bomb fallout is nearly identical to the fission products given off by ordinary nuclear power plants during their "routine emissions" and accidents.) Atomic-industry workers, long assured that they were being protected despite lax radiation-exposure standards, are dying of cancer at a higher-than-normal rate. Dr. Ernest Sternglass has found increased levels of cancer downwind from nuclear power plants and other atomic facilities.

Long-lived radioactive poisons, such as plutonium-239, which remains deadly for at least a quarter million years, will recirculate through the world's ecosystems for thousands of generations, with the potential for causing fatal cancers whenever they are ingested. The more nuclear power plants we build, the more fission products will be released to the environment; the more radioactive waste will be stockpiled; the more likely it is that there will be a serious nuclear accident killing tens of thousands of people and irradiating hundreds of square miles of land.

The time to stop nuclear power is right now.

George R. Zachar

There's No Such Thing as a "Peaceful Atom"

Jerry Elmer

In recent years demonstrations commemorating the anniversaries of the atomic bombings of Hiroshima and Nagasaki have called not only for nuclear disarmament but also for an end to nuclear power plants. In response to such protests seeking to link these two issues, utility companies involved in nuclear plant construction have been quick to state that connections between the two issues do not exist. "We feel there is no more connection between nuclear power and the bombing of Hiroshima than there is between electricity and the electric chair," said one spokesperson for New England Power Company (NEPCO). "It's like comparing air travel with a B-52 bomber," said another NEPCO spokesperson.

When utility companies say that nuclear power plants cannot explode like an atomic bomb, they are correct; such an explosion is a physical impossibility for conventional nuclear power plants. Nevertheless, there are inescapable links between the two issues with which many people are not familiar.

First, the nuclear reaction that takes place in nuclear power plants is identical to the nuclear reaction that took place in the Hiroshima bomb—the splitting of the uranium-235 atom. Thus, the radioactive "daughter" elements produced in the two reactions—including krypton-85, xenon-133, strontium-90, and cesium-137—are identical. Also, the dangerous radiation produced is the same.

Radiation from this reaction does not directly penetrate the walls of the nuclear power plant because of the heavy shielding. However, many of the daughter elements of the fission process do escape from the reactor—either through routine emissions or, as in the case of Three Mile Island, by accident. Once unleashed in the environment, these daughter elements continue to decay by emitting radiation. At Hiroshima, enormous levels of radiation contributed to the deaths of tens of thousands of people immediately and produced painful, lingering deaths from radiation sickness in countless others. At the much lower levels at which such radiation is emitted from nuclear power plants in the United States, such radiation has been linked to the dramatically increased incidence of cancer, leukemia, and terminal gastrointestinal disorders. A 1973 study by the National Academy of Sciences concludes that exposure of the entire U.S. population to what the government calls acceptable, routine emissions could cause 15,000 additional cases of cancer *per year*.

One of the strongest links between nuclear weapons and nuclear power is the relationship between the civilian nuclear industry and worldwide proliferation of nuclear weapons. Simply stated, nuclear power plants spread both the technological knowhow and the raw materials needed to build atomic bombs. The only missing element is the costly reprocessing technology needed to separate plutonium and fissionable uranium, the leftovers from the

Thermonuclear detonation. Credit: Lookout Mountain Air Force Station.

Weapons Production Facility. Credit K.C. Bendix Corporation Photo.

nuclear reaction that are the most suitable for weapons
production, from the spent (used) fuel. Several nations
have experimental or research reprocessing plants in operation. Others are trying to obtain the technology.

On May 18, 1974, India became the sixth nation in the
world to explode an atomic bomb; the bomb was built from
materials supplied to India for its program of civilian
atomic power. In addition to the six nations now possessing
nuclear weapons, 24 other countries have the technological expertise and the nuclear fuel to build bombs at any
time. Eight of the 24—including Brazil, South Africa,
Egypt, Israel, and Pakistan—are not signers of the Nuclear
Nonproliferation Treaty of 1970. Each of the 24 achieved
its nuclear potential from materials supplied for its civilian
nuclear power program. Reprocessing equipment is all
that separates these nations from those that belong to the
nuclear-weapons club.

The crucial element in this connection is plutonium-239:
the same plutonium manufactured by nuclear power
plants is the fuel used in most atomic bombs manufactured
today.

Plutonium is widely regarded as the most toxic substance
known. Microgram quantities (one-millionth of a gram) absorbed in the skin will cause skin cancer; deposited in the
bone it causes bone cancer (biologically, plutonium is a
bone-seeker). A piece of plutonium the size of a grain of
pollen will, if inhaled, almost invariably produce lung cancer. If it were possible to distribute one pound of
plutonium evenly, it would be enough to kill every human
being on Earth.

Plutonium is one of the elements produced in civilian
nuclear power plants. A typical 1000-megawatt light-water
reactor produces 400 to 600 pounds of plutonium each
year. If the nuclear industry has its way, the United States
will produce some 280,000 pounds of plutonium by 1985.
Plutonium has a half-life of 24,000 years; that is, after 24,-

000 years it has lost only one-half of its radioactivity. Because of its extreme toxicity, every ounce of plutonium will have to be kept safe from accidental leakage, earthquakes, sabotage, terrorism, and acts of God for hundreds of thousands of years. The leakage of even a fraction of 1 percent of our plutonium could have disastrous consequences. The mediocre record of the nuclear industry to date in safeguarding against environmental leakage of radioactive materials gives one considerable cause for concern.

Imposing safeguards on exported nuclear material, something periodically debated in Congress, is not likely to be successful. India's atomic bomb was made from plutonium extracted from spent nuclear fuel supplied by Canada—at a time when Canada imposed stricter safeguards on nuclear material than the United States has yet to impose. The Canadians were convinced that such diversion by India to make a bomb was impossible. Nevertheless, it happened.

To date, the United States and other nuclear powers have exported 111 nuclear power plants to at least 25 recipient nations. The United States, having exported 57 of those reactors, accounts for more than half of all the world's reactor exports. Each of the 111 exported reactors produces enough plutonium to build an atomic bomb approximately every two weeks. Naturally, with such tremendously increased quantities of fissionable material around, the danger of nuclear fuel falling into the hands of private terrorist organizations is also significantly increased. Many scientists fear that if present trends continue, nuclear weapons may become for the next generation of terrorists what machine guns are today.

Already there is enough plutonium circulating that it is apparently impossible to keep track of it all. On December 24, 1974, the *New York Times* reported that as much as 60 pounds of plutonium was unaccounted for at the Cimarron nuclear power plant in Oklahoma. Another recent report

of missing fissionable fuel was in the *Times* on March 24,
1978, saying that 202 pounds of highly enriched, bomb-
grade uranium—enough for at least 10 bombs—was missing from a nuclear facility in Apollo, Pennsylvania. No one
seems to know if this material was stolen in one big robbery
or pilfered by staff over a period of time; no one knows if
this discrepancy resulted from a "bookkeeping" error or if
the fissionable material has already found its way into
bombs. On August 6, 1977, the *Norwich* [Connecticut]
Bulletin reported that 160 pounds of uranium was unaccounted for from Connecticut's four nuclear power plants
over the last 18 years. (Governor Ella Grasso's press secretary issued a statement that there was no cause for alarm,
since this represented only 0.5 percent of the nuclear fuel
used in Connecticut over that period and thus was well
within U.S. government guidelines!)

Contrary to claims by the nuclear industry, building
bombs is not extraordinarily difficult once the necessary
fuel has been obtained. John Phillips, a Princeton under-
graduate who ran a pizza business, recently wrote a paper
on how to construct a homemade bomb and received inquiries about his design from two foreign powers. Dmitri
Rotow, a Harvard student, designed a series of 22 atomic
bombs in five months; two bomb designers for the U.S.
government called some of his designs "highly credible."

Many Americans feel that nuclear proliferation is dangerous because, they believe, leaders of Third World countries are inherently less intelligent or responsible than
leaders of big powers. It is useful to recall that the failure
of the Nonproliferation Treaty of 1970 (NPT) is largely
attributable to the nuclear superpowers rather than to
smaller, nonnuclear nations. The NPT called for a two-part
bargain: the nuclear powers promised to work actively
toward nuclear disarmament, while nonnuclear powers
promised to forgo the option of "going nuclear." In the
years since the NPT was signed, the world's nuclear pow-

ers have continued designing and testing a wide range of new nuclear weapons (such as the neutron bomb) and new delivery systems (such as the Trident submarine, MIRV missile, and cruise missile). The image of nuclear superpowers, each with the capacity to wipe out all life on this planet, preaching nonproliferation to nonnuclear Third World countries would be comical if it were not so tragic. In this context, the pressure felt by many nations to "go nuclear" is understandable.

Nevertheless, continued proliferation of nuclear weapons does pose many hazards. The more nations that have nuclear weapons, the more likely that a local war (such as between India and Pakistan or in the Middle East) could cross the brink from conventional to nuclear. Such a confrontation could, in turn, draw in the superpowers, with their vast nuclear arsenals. And, of course, the more nuclear material there is circulating in more countries, the greater the chance of fissionable material falling into the hands of private terrorist organizations.

Despite the obvious problems created by export of nuclear power plants, there are several reasons why such export is not likely to stop. Two U.S. corporations, General Electric and Westinghouse, account for 50 percent of the total international sales of nuclear power plants. These companies count on such foreign sales for a significant portion of their profits. Furthermore, in order to bring the per-unit cost of power plants down for the domestic market, these corporations try to increase total sales volume by increasing foreign sales. In addition, nuclear exports, now amounting to over a billion dollars per year, are important to the United States balance of trade.

Another important connection between nuclear weapons and nuclear power is that new technologies developed to help the civilian nuclear industry can also help countries (or private terrorist organizations) to develop nuclear capability for nonpeaceful purposes. The best-known example of this is the development of the "breeder reactor," designed (but not yet produced commercially in the U.S.) to help the nuclear industry overcome the impending shortage of uranium as nuclear fuel. The breeder reactor would use plutonium as fuel and would actually produce more plutonium than it consumes, yielding a virtually limitless supply of fuel. Of course, it would also contribute enormously to the potential proliferation of nuclear weapons by increasing the amount of bomb fuel in circulation.

A lesser-known but very important example of technology aiding proliferation is the current development of laser technology for uranium enrichment. When it is mined from the ground, natural uranium contains two isotopes: uranium-235, which is fissionable and can be used either in power plants or in bombs, and uranium-238, which is not fissionable. When mined, natural uranium consists of 0.7 percent U-235 and 99.3 percent U-238. Before it can be used, the uranium must be enriched—that is, have the level of fissionable U-235 increased to at least 3 percent for power plants or 20 percent (minimum) for bombs.

The classical means of uranium enrichment is called gaseous diffusion—a gassified uranium compound (uranium hexafluoride) is passed through a series of filters that separate the two isotopes. This process is vastly expensive and can never be done secretly: gaseous diffusion plants cost nearly $3 billion and take up acres of land. The astronomical cost of uranium enrichment has been one of the primary obstacles to many small countries' building their own bombs; the superpowers had the monopoly on uranium enrichment. The high cost of uranium enrichment has also been an important factor in keeping the cost of nuclear-produced electricity high.

As a result, the American nuclear industry has been searching for a way to bring down the high cost of enrichment. That way may now have been found; the new method—still in experimental stages but demonstrably workable—utilizes laser beams. Scientists take advantage of the different resonating frequencies of U-235 and U-238 by directing a finely tuned laser at the unenriched uranium (also in the form of "hex" gas) to excite the molecules of one isotope and not the other. From there, separation is a quick and relatively cheap process.

So promising is this laser enrichment of uranium that some scientists fear that in the future pilferage or theft of fissionable material will cease to be a threat—because manufacture of fissionable material will be so easy! One of the major obstacles to the uncontrolled spread of nuclear weapons—to other countries or private terrorist organizations—is being eroded, as a result of technological developments for the civilian nuclear power industry.

Since the atomic bombings of Hiroshima and Nagasaki, many peace-loving people the world over have fervently hoped that it would be possible to use nuclear energy for exclusively peaceful purposes—that is, that nuclear energy for civilian purposes could be separated from nuclear power for military purposes. This has not proved possible. It is now clear that many of the same health, genetic, and environmental problems created by nuclear bombs are also inherent in the civilian nuclear industry. It is also clear that the goal of nonproliferation of nuclear weapons is virtually unattainable so long as there is a civilian nuclear power industry.

Waving Goodbye to
the Bill of Rights

Donna Warnock

In August 1974 news came from Texas that the state police had admitted compiling a dossier on Robert Pomeroy, an airline pilot and head of Citizens Association for Sound Energy (CASE). According to the file: ". . . subject is using CASE as a front group—possibly for a Ralph Nader action."

Word about the Pomeroy affair traveled fast. That such surveillance was in force came as no surprise to a country whose attention was riveted that year to Watergate and the Nixon resignation. But was Pomeroy-type surveillance simply a Watergate phenomenon? Or was it further fallout from our nation's budding nuclear power program? A close look at the issue reveals that nuclear power requires diligent surveillance and that intelligence operations are virtually uncontrollable—particularly in the hands of private industry and government agencies whose actions are rarely subject to public scrutiny.

The first and one of the most important lines of defense against groups which might attempt to illegally acquire special nuclear materials to make a weapon is timely and in-depth intelligence. Such intelligence may involve electronic and other means of surveillance, but

its most important aspect is infiltration of the groups
themselves. . . . There must also be an ongoing analysis
of the attitudes of the people in the plant and the community around the plant.

So concluded the 1974 Rosenbaum Report, prepared for
the Atomic Energy Commission (AEC)—a document the
government had tried, without success, to keep from public disclosure. The FBI is required to conduct investigations of nuclear theft, diversion, or sabotage under the
1954 Atomic Energy Act. The bureau began to increase its
investigations under the Act in 1974 and announced in its
1974 annual report that it expected the upward trend to
continue.

That year also saw the mysterious death of plutonium
worker Karen Silkwood. The circumstances were suspect.
Silkwood might have been the target of surveillance and
murder, but a company and government coverup kept the
Silkwood family from gaining headway in their search for
the facts.

Silkwood's death brought the nuclear surveillance issue
to the public. Too many questions remained unanswered.
Too much evidence pointed to the possibility that the Silkwood case was just the tip of the iceberg.

Then, in 1975, Russell Ayres' comprehensive and com-
pelling review of the civil liberties problems attendant to
nuclear power, published in the *Harvard Civil Rights–
Civil Liberties Law Review*, concluded that plutonium
"provides the first rational justification for widespread surveillance of the civilian population."

The reasoning was clear: government experts estimate
that an accident at a major nuclear power plant could kill
countless thousands and contaminate an area the size of
Pennsylvania for thousands of years. Harm from an act of
malice such as a terrorist action could be even greater.
Thus, protection from thefts or sabotage is critical, particu-

larly since over 230 bomb threats have already been aimed at domestic nuclear power facilities, and explosives have been found at U.S. reactors on four occasions, according to the Nuclear Regulatory Commission (NRC). One was actually detonated at the Trojan nuclear reactor in Oregon but caused only minor damage to the plant's visitor center. Yet security at nuclear facilities is so poor that over 30,000 pounds of nuclear materials are currently missing from civilian and military facilities. It takes less than 30 pounds of highly enriched uranium to make an atom bomb. The threat of basement A-bombs has become all too real.

Ayres' warning began to hit home. A bill was introduced in the Virginia State Legislature in 1975 that would permit the Virginia Electric & Power Company (VEPCO) to establish its own police force with statewide arrest powers and access to confidential state and local police records. VEPCO claimed the authority was required to meet the Atomic Energy Commission's nuclear-safeguards regulations. The bill failed. A similar bill was introduced in the Washington State Legislature in 1979 and it also failed.

Disconcerting news continued to come to light. *Counterspy* magazine revealed in 1975 that the Atomic Industrial Forum, the nuclear industry trade association, and the New York consulting firm Charles Yulish Associates were regularly providing the utility industry with profiles and status reports on people and organizations active in the fight against nuclear power. Environmental Action, the Union of Concerned Scientists, the Sierra Club, Friends of the Earth, the Environmental Policy Center, Another Mother for Peace, and Ralph Nader were all among the targets.

Increasingly, other antinuclear groups around the country began to suspect that they too were the victims of surveillance. The Abalone Alliance in California, whose activities have been monitored by the FBI since 1976, discovered that it had been infiltrated by two sheriff's deputies during its occupation of the Diablo Canyon nuclear plant in August 1977. The two were the only occupiers who suggested using violence. One of the infiltrators attempted to sabotage the group's legal strategy for charges stemming from its trespass arrests. Defense attorneys contended that the group's attorney-client privilege had been violated. At this writing the case is pending before the California Supreme Court.

Filming of protestors has become regular fare at antinuclear events. Los Angeles police were asked by the L.A. City Council to stop filming individuals as they testified against a proposed nuclear plant before the Council in February 1978. Seventeen of those who testified filed suit against the police and are seeking an injunction to prevent future police photographing.

In New Hampshire, state police parked outside the Clamshell Alliance headquarters for a short period in 1977 and made videotapes of visitor traffic.

Charges of potential violence were used by the Philadelphia Electric Company (PECO) as an excuse for photographing demonstrators, a PECO employee said in a televised interview. This prompted a group of nuclear opponents, ratepayers, and stockholders to ask the Pennsylvania Public Utility Commission in August 1978 to investigate PECO's surveillance activities.

In September 1977 the *Atlanta Journal* reported the great lengths to which the Georgia Power Company had gone to spy on opponents. In 1973 the utility started its special surveillance unit, which operates on an annual budget of $750,000. Its nine investigators were equipped with walkie-talkies, wiretapping and videotape equipment, and James Bond-type cars with headlights that change configuration at the flip of a switch to confuse people they are following. Both Georgia Power and Pacific Gas and Electric use the information services of Research West, a firm that refused to show its files on nuclear opponents to Con-

gress because it claimed doing so would force the company
to "go out of business."

In February 1979 the NRC denied the Houston Chapter
of the National Lawyers Guild intervention status against
the Allens Creek nuclear facility because the guild
wouldn't give the NRC the names and addresses of its
members—ironic, since the group's petition to intervene
was based on civil liberties concerns.

While all this was going on, the evidence in the Silkwood
case was piling up. Years of careful investigation proved
that the suspicions of wrongdoing were indeed on target.

In 1974 Karen Silkwood had protested health and safety
practices where she worked at the Kerr-McGee Corporation in Oklahoma. She was harassed, possibly wiretapped,
and mysteriously contaminated with plutonium found in
her home. The highest levels of contamination were found
on the bologna and cheese in her refrigerator.

Silkwood was killed on the night of November 13, 1974,
when the car she was driving crashed. The official Oklahoma State Police accident report concluded that Silkwood lost control of her car after she fell asleep at the
wheel. However, other evidence indicates that she may
have been murdered. A private accident investigator hired
by her union, the Oil, Chemical, and Atomic Workers
(OCAW), found chemical and physical evidence that her
car was forced off the road by another vehicle.

Silkwood had been on her way to present a *New York
Times* reporter and a representative of the OCAW with
documents showing that Kerr-McGee was falsifying qual-
ity-control records in the production of plutonium fuel
rods. The evidence a witness saw her take to her car vanished from the scene of the accident. Other witnesses ar-
riving soon after the crash said they saw documents with
Kerr-McGee's name on them blowing around the car.

Roy King, Kerr-McGee's personnel manager, has sworn
under oath that he received a call the night of the crash

from a state police officer, Rick Fagan, who said he had picked up company papers at the crash site and had placed them in the wrecked car. The two men made an appointment for the next morning to go to the car together to retrieve the papers for Kerr-McGee. However, according to King's statement, Officer Fagan stopped by the next morning to say that the documents had already been removed.

According to FBI and other government accounts of the events following the crash, at about one A.M., soon after the wrecked car had been towed from the accident site to a garage in nearby Crescent, Officer Fagan, four Kerr-McGee officials, and two representatives from the Atomic Energy Commission visited the car. They said their purpose was to check the wreck for radiation contamination. Whether these officials saw or removed the papers is still unknown because their testimony was ruled irrelevant in the first trial concerning Silkwood's contamination with plutonium. However, their testimony would probably be heard if the Silkwood family is successful in bringing a second lawsuit to court concerning civil rights violations.

Regardless of what happened to the papers—and that is still a mystery—they were gone by the time representatives from the Oil, Chemical, and Atomic Workers inspected the car the following morning. They found only a few papers, none bearing Kerr-McGee's name.

In May 1979 the Kerr-McGee Corporation was found guilty of negligence leading to plutonium contamination in Silkwood's home the week prior to her death. The judgment, which resulted from a lawsuit by Silkwood's family, ordered the company to pay $10,505,000 in damages. In addition to the very large penalty, the case is significant because the jury's decision asserts "the theory of strict liability" whereby nuclear facilities have a greater responsibility than do other enterprises to safeguard materials in their plants because of the extraordinary threat they pose to life and property. If these substances are released, the jury ruled, nuclear facilities should be held financially responsible for the damage or injury they cause. Furthermore, the decision states, Kerr-McGee is responsible for Karen Silkwood's contamination even if the company was in "substantial compliance" with federal regulations in operating the plant. If upheld on appeal the decision broadens the definition of bodily injury to include damage to tissue, bones, or cells that may be undetectable initially.

The second lawsuit by the Silkwood family charges individual members of the board of directors and management of Kerr-McGee with conspiring to violate Silkwood's civil rights. The suit also charges four FBI operatives with covering up the conspiracy. These civil rights counts were dismissed in September 1978, when the court ruled that existing civil rights legislation outlaws private conspiracies only against racial or ethnic minorities. The Silkwoods are appealing the decision.

Much new evidence has come out in the court documents generated by the Silkwood case. An Oklahoma City Police Department (OCPD) "associate," Steve Campbell, developed a friendship with Silkwood's boyfriend Drew Stephens, in order to gain access to "intelligence data" and confidential diaries. This information was bought by Kerr-McGee. A former Kerr-McGee secretary has testified that while she worked for the Oklahoma Police Department, OCPD officers engaged in illegal wiretapping, breaking and entering, and electronic surveillance, and that she had typed transcripts of the wiretaps. Silkwood investigators also found that the OCPD had been gathering information on Silkwood while she was alive and routinely providing it to James Reading, Kerr-McGee's security chief. Subpoenaed court documents further revealed that a paid Kerr-McGee undercover agent had entered Silkwood's home and photographed the notes she had collected on health, safety, and plutonium fuel-rod specifications.

Marlette. Copyright The Charlotte Observer. King Features Syndicate.

Among the FBI documents released in the Silkwood case was a bureau report revealing that FBI informant Jacque Srouji wrote a book on nuclear power to develop FBI contacts in the field. During 1976 hearings on the Silkwood case before Congress, the FBI Deputy Assistant Director, James Adams, was asked whether the FBI investigates antinuclear groups. Adams responded, "The Communist Party of the United States, which is dominated and controlled by the Soviet Union . . . has as a program to try to discourage the use of nuclear energy in the United States." Therefore, he said, the FBI has an interest in other groups with such a position and such "information may be developed and maintained."

The U.S. government has routinely been involved at many levels in collecting information on nuclear critics. The FBI provides both the Department of Energy and the Nuclear Regulatory Commission with advance warnings about individuals and groups that it believes may pose

threats to nuclear facilities. In addition, FBI files include
the results of over 2000 investigations (including an un-
specified number of employee checks) per year for the
Department of Energy and the NRC. Every FBI Field
Office has one or more special agents who are assigned to
act as liaison officers for nuclear facilities. The U.S. General
Accounting Office has found that utility employees were
regularly used as "confidential informants" in FBI investigations.

The private Law Enforcement Intelligence Unit (LEIU)
facilitates the exchange of intelligence dossiers among its
225 local police unit members. Utility officials have also
been privy to LEIU meetings. Federal laws defining what
private companies can and can't do to collect information
are sketchy at best.

The National Environmental Policy Act (NEPA), passed
in 1969, requires that all federal agencies assess the environmental impacts of their actions and various alternative
options. NEPA clearly can be interpreted as requiring that
the civil liberties impacts of agency actions, including the
use of nuclear energy, be assessed. Yet no nuclear environmental impact statement has ever discussed civil liberties
repercussions, and this application of the law has not been
tested in court.

The prospect of reprocessing nuclear fuel to extract
plutonium for use in nuclear breeder reactors (which then
produce more plutonium than they consume) is particularly ominous, since plutonium is prime bomb material.
Less than 10 pounds are needed to create a crude nuclear
weapon—and once that is in hand it can easily be fashioned
into an explosive. Even the simple dispersion of plutonium
alone—not in a bomb—could cause a large number of
deaths in a major city. Properly distributed, just one pound
could kill everyone on Earth.

Commercial reprocessing has been prohibited by the
U.S. government, at least temporarily, due to concern that

it would lead to production of nuclear weapons by nations that do not have them—not to mention the threat of nuclear terrorism. But diminishing quantities of uranium to fuel power reactors give nuclear proponents a rationale to continue pushing hard for reprocessing.

It is important to point out that the government's moratorium on plutonium reprocessing has not eliminated the possibility of plutonium theft. Each 1000-megawatt reactor produces 400 to 500 pounds of plutonium in a year, enough to make 40 or more nuclear bombs. It is possible to separate this plutonium from spent reactor fuel in a laboratory set up for such purposes, although it would require sophisticated and expensive equipment. Reprocessing on a large scale would create a plutonium economy, with millions of tons of it "routinely" being shipped around the world. Once separated from spent nuclear reactor fuel, plutonium would be a prime target for terrorists and nations trying to obtain atomic weapons because it could readily be fashioned into weapons, the costly and difficult reprocessing step having been done for them.

There is no guarantee that the plutonium already missing has not found its way into the wrong hands. For example, forty pounds of plutonium is still unaccounted for at the Kerr-McGee plant where Karen Silkwood worked. The company claimed that plutonium was lost in the plant's piping, although repeated pipe flushings failed to turn up the missing material. A former supervisor at the same plant testified under oath that when he worked at another Kerr-McGee facility, he was twice asked to divert plutonium from government stockpiles to Kerr-McGee stockpiles.

It seems likely that a black market trade in plutonium will develop, particularly if we proceed with the recycling of plutonium. Former Atomic Energy Commissioner Clarence Larson has predicted:

Once special nuclear material [plutonium or weapons-grade uranium] is successfully stolen in small and possibly economically acceptable quantities, a supply-stimulated market for such illicit material is bound to develop. And such a market can be surely expected to grow once the source of supply has been identified. As the market grows, the number and size of thefts can be expected to grow with it. . . . Such theft would quickly lead to serious economic burdens to the industry, and a threat to the national security.

In February 1977 a special federal task force reached the frightening conclusion that the government "could not with a high degree of assurance, prevent the diversion of nuclear materials by an insider." For this reason, atomic workers face many infringements on their civil liberties—including background checks, security clearances, behavioral analyses, physical searches, questioning, lie-detector tests, and even detention in the event some nuclear material is believed to be missing from a facility. Hundreds of thousands of civilian workers are affected.

Many government and nongovernment studies have shown that nuclear facilities are, in fact, far from adequately protected. They have weak physical barriers, ineffective alarm systems, and poor guard patrols. A report issued in April 1977 by the U.S. General Accounting Office, after visits to several nuclear power plants, states: "[D]uring our site visits . . . inspectors observed such weaknesses as guards not responding to an alarm and an unlocked exterior door which permitted unrestricted access to the control room."

Sheer luck has prevented nuclear catastrophe from occurring to date. However, it is doubtful that this good luck will continue as the number of nuclear plants worldwide grows, as the number of nations with nuclear capability increases, as the number of nuclear shipments rises, and as

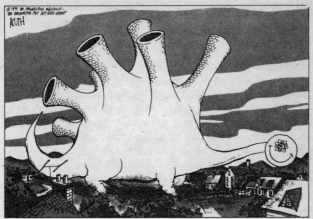

Tony Auth. © 1979, The Philadelphia Inquirer. The Washington Post Writers Group.

the number of people with basic technical skills in nuclear science expands.

The only apparent solution for a future of nuclear power is to greatly tighten up those existing safeguards and add new ones. According to a study prepared for the Nuclear Regulatory Commission by Stanford University law professor John Barton, in the case of a plutonium theft, emergency evacuation might be necessary, cordoning of large areas imposed, homes searched, property confiscated, and personal privacy violated. In addition, the government might resort to the seizure, detention, and torture of dissidents, the study predicted. Though torture is illegal and unconstitutional, Barton and others are not certain that the law would deter such actions during a perceived nuclear emergency, since, as Russell Ayres points out, we have not yet had to ask "questions like whether it is better to torture a suspected terrorist than let a city go up in flames." Such an incident could create a public reaction supporting torture and other serious civil liberties infringements.

In the case of an emergency involving dispersion of radioactive contamination, the Federal Response Plan for Peacetime Nuclear Emergencies (which is coordinated by several dozen federal agencies) indicates that the basic political, social, and economic systems and values of the affected area could be severely disrupted. The exercise of constitutional rights and liberties might be curtailed, representative constitutional government abandoned, arbitrary police powers practiced, and the free economy and private operation of industry restricted.

In welcoming the use of nuclear energy, we are waving goodbye to the Bill of Rights. What "rights" do we really have if the threat of martial law and torture hovers above us, and both the private and public sectors resort to widespread surveillance? The writing is on the wall; we can have nuclear power or we can have civil liberties. But we can't have both. The home of nuclear power can never be called "the land of the free."

Who Got Us into This Mess?

Mark Hertsgaard

Barely 48 hours had passed since Wednesday's pre-dawn accident at the Three Mile Island nuclear power plant. Confusion and rumors were everywhere. Initial news reports were vague about what was actually going on inside the sullen hulk of steel and concrete that punctuated the skyline of Middletown, Pennsylvania. The operators of the plant, the Metropolitan Edison electric utility company, had put their best foot forward in assuring everyone that the situation was under control. That foot ended up in Met Ed's mouth by Friday morning, as reports filtered out that a large hydrogen bubble had formed inside the reactor. The U.S. Nuclear Regulatory Commission (NRC) admitted that the ultimate nightmare of nuclear power technology —a core meltdown—had become a dangerously real possibility.

That very Friday morning, Westinghouse executive Leo Yochum was being interviewed in his Pittsburgh office about the nuclear power business. As the number-three man at the atomic industry's leading firm, Yochum was well acquainted with the subject matter. As the head of corporate relations for Westinghouse, he keeps in close touch with investment banks, insurance companies, and other arms of Wall Street power.

When asked if Westinghouse, the most nuclear-committed company among the reactor manufacturers, was having second thoughts about staying in the business, Yochum boomed back, "No, why should we? Our nuclear work has been very profitable for us."

After acknowledging his disappointment with the industry's dismal performance since the bottom dropped out of the reactor market in 1975, Yochum was still buoyantly confident about the future. He sat roughly 200 miles from the stricken reactor that threatened thousands of Pennsylvanians and summed up the nuclear industry's faith with almost evangelical fervor.

"I just don't understand this talk about nuclear being dead. The market is going to return. After all, there *is* a nuclear imperative for this country. We know it, Wall Street knows it, and we're prepared to meet it."

Mr. Yochum's comments reveal much about the collage of corporate power that would like to bring nuclear electricity to every community in America. The first question that most people ask about nuclear power is "How safe is it?" The first question a nuclear company must ask is "How profitable is it?"

Westinghouse, General Electric, Combustion-Engineering, and the Babcock and Wilcox subsidiary of J. Ray McDermott are the Big Four reactor manufacturers, or vendors, who monopolize the U.S. nuclear reactor market. They are now trying to decide how much longer to wait for the lifeless domestic market to revive before diverting their investment dollars from nuclear plants into more profitable enterprises.

The nation's 200-plus large electric utilities, along with dozens of rural electric cooperatives and public power systems, provided a brisk business for the nuclear vendors until 1975. Since then, sluggish demand for electricity has cast a shadow over the entire utility industry and has led to more reactor cancellations than orders.

The three architect-engineering firms who have constructed most of the nation's (and the world's) reactors—Bechtel, Stone and Webster, and United Engineers and Constructors—lament the loss of the plumper pocketbooks nuclear-plant manufacturing once brought them.

Kerr-McGee, United Nuclear, Gulf Oil, and the other mining and energy companies that dominate the uranium business are undoubtedly elated over the $42 that a pound of uranium now fetches, but they worry about the pending outcome of Westinghouse's lawsuit accusing them of illegally driving up uranium prices.

Morgan Stanley, Chase Manhattan, and the other Wall Street interests upon whose investor confidence and massive funds of risk capital the utilities and vendors depend are convinced that full-scale pursuit of the nuclear option is critical to enhancing the American economy's ability to respond flexibly to the future whims of international energy politics. But they fret that growing antinuclear sentiment may force further government restrictions that will make nuclear power unprofitable.

Behind all the patriotic rhetoric, companies ultimately evaluate nuclear power on the basis of cold dollars and cents. The nuclear industry's worry about Three Mile Island is not that it vindicates environmentalists' charges of unacceptable nuclear risks but that the inevitable addition of further government regulations will price nuclear power out of the market.

There is another lesson in Mr. Yochum's optimism: neither the Three Mile Island accident nor the growing public concern about nuclear's health and safety will sway the nuclear power industry from its single-minded mission to spread nuclear energy across America. After investing billions of dollars in the business, and with billions more in potential future profits, the industry is hardly ready to give up because a cooling pump malfunctioned. The struggle over nuclear power is far from over.

IN THE BEGINNING

Yochum's devotion to nuclear power is not surprising, considering how long and deeply his company has been involved with it. Westinghouse was there at the very beginning, as one of the main contractors for the government's World War II Manhattan Project—the project to develop the atomic bomb. That effort brought together many of the scientists and companies that would later populate the commercial nuclear power industry.

Operating under intense wartime pressures, the Manhattan Project legitimized the extraordinary government-industrial partnership that still shapes U.S. nuclear power development. It bequeathed to the commercial program the advantages and disadvantages of the secrecy, subsidies, and military preoccupations that characterized it during the nuclear race against the Nazis.

For example, the same top-secret security system developed to guard against Nazi spying in the early 1940s would again be invoked during the Nevada bomb tests of the 1950s, when the army used soldiers as human guinea pigs for testing their reactions to atomic blasts. Cloaking their action behind the slippery concept of "the national security," the government would cover up the story and later deny it ever happened. Washington would lie again in the 1960s to suppress official documentation of the hazards of radiation.

The same federal treasury that paid for the uranium enrichment and plutonium facilities used to make the Hiroshima and Nagasaki bombs would be relied upon countless times over in the next 30 years. Reactor research and development, accident insurance, fuel preparation, and reactor export subsidies all would be paid for with taxpayers' money.

The 1946 Atomic Energy Act institutionalized the government's role in nuclear development. It shifted control over the technology from the army to the newly created civilian-staffed Atomic Energy Commission. The AEC would be the government's main tool in commercializing nuclear power. Along with the Department of Defense, it would also oversee the weapons-production program.

The act granted the AEC total power over all nuclear facilities, weapons, and future nuclear-related production. This unusual step was not taken just for security reasons. According to the AEC's official history, what stirred Congress "to advocate unprecedented government intervention in the economic process was the anticipation of substantial if not spectacular innovations in nonmilitary uses of atomic energy."

The extraordinarily broad powers granted to the AEC were paralleled in a new Joint Committee on Atomic Energy (JCAE) set up by Congress, which was charged with serving as the "people's watchdog" over the nuclear program. It evolved into one of the most powerful forces on Capitol Hill. Critics would soon charge that it ignored im-

By permission of Bill Sanders. The Milwaukee Journal, 1979. Courtesy of Field Newspaper Syndicate.

"Why he doesn't trust us is hard to see!" said Tweedledum to Tweedledee

portant health and safety questions in order to protect the
nuclear industry's business prospects.

The 1946 act, the cornerstone of all subsequent nuclear
legislation, outlined the contradictory institutional responsibilities that continue to undermine meaningful federal
regulation of nuclear power. The act ordered both the
AEC and the JCAE to develop the military and commercial applications of nuclear energy so as to "defend the
common security and enhance the general welfare."

They were also charged with the responsibility to regulate nuclear power. The tremendous access of corporate
interests to the levers of government power, amplified beyond normal levels by the veil of secrecy demanded by the
"national security," stimulated an ingrained bias within
both bodies toward their promotional, rather than regulatory, functions. This has marred government policy ever
since.

Until 1952 most of the AEC's time and money went into
assembling the world's most terrifying nuclear arsenal. But
the dream of commercializing nuclear power was not forgotten. And as other nations initiated their civilian nuclear
programs, the U.S. drive to stay on top of nuclear technology demanded an accelerated commercialization program.

The impetus for commercialization came most strongly
from the joint committee. In the summer of 1953 the JCAE
transformed Admiral Hyman Rickover's plan to develop a
nuclear aircraft carrier into a project that in four years'
time would produce the first U.S. nuclear power plant at
Shippingport, Pennsylvania. The selection of the light-
water reactor (LWR) technology guaranteed both the
LWR design and Westinghouse, who developed it, a bright
future in the nuclear business.

The joint committee made the Shippingport decision
following hearings that concluded that private industry
could make nuclear energy profitably, provided the proper

type and scale of government assistance was forthcoming. Investment banker Lewis Strauss, then the AEC chairman, summarized the corporations' perspective on a future nuclear power industry:

> Executives in private industry believed that there ought to be assurances that the government was willing to allow a private atomic energy industry to develop, that risk-taking would be compensated by profits for success and financial allowances made for failures; and that regulations would be established with progress and profit in mind as well as safety and security. Amendment of the Atomic Energy Law was, of course, a prerequisite before private industry could own or even lease fissionable material for fuel.

The law was so amended the very next year. The 1954 Atomic Energy Act signaled how serious the government was about developing commercial nuclear power. It allowed corporate ownership and operation of nuclear reactors, as well as access to enriched uranium fuel.

The government's committment was again put to the test during the AEC's "Power Reactor Demonstration Program," which lasted from 1955 to 1959. The program's purpose was to introduce utilities to the dozen or so companies then interested in manufacturing nuclear reactors. Results of the subsidized joint construction projects were often frustratingly ambiguous or even negative, as numerous designs were tested and discarded.

The companies almost seemed to play hard to get. In 1957 the Atomic Industrial Forum, the industry's newly formed trade association, published a survey that cited a lack of economic incentive and insurance protection as the two major reasons why the private sector still hung back from a full-fledged drive to commercialize nuclear power.

The government's 1957 Brookhaven report had sent

shivers up utility executives' spines. It stated that, in a worst-case accident, thousands of people might die and property damage could run into the billions. Private insurance companies refused to accept such a risk, and utilities would not buy nuclear plants without protection against those risks. As the price for continuing in the nuclear business, the companies convinced Congress to pass the Price-Anderson Act—a law limiting the firms' financial liability for a nuclear accident to a meager $560 million.

The Eisenhower administration had two international counterparts to the domestic Power Reactor Demonstration Program—the Atoms for Peace and EURATOM programs. Although a U.S. nuclear monopoly was no longer possible when Ike delivered his famous Atoms for Peace speech at the U.N. in 1953, the United States still hoped to control the future international flow of nuclear merchandise.

Both programs offered foreign nations money and technical assistance to construct U.S.-designed reactors. The Atoms for Peace reactors were usually financed by the Export-Import Bank, a government agency set up to subsidize U.S. exports with low-interest loans to foreign nations. These deals served partly as tools in the Cold War effort to strengthen diplomatic ties throughout the world. The EURATOM program, which subsidized the introduction of the American light-water technology into Europe in the late 1950s, was an opportunity to, in the words of the joint committee, "demonstrate U.S. leadership in atomic energy."

These two programs stimulated the U.S. nuclear industry with millions of dollars in business and invaluable practical experience at a time when the domestic program was still getting untracked. More important, subsidizing the foreign adoption of American light-water reactor technology assured that U.S. firms would initially dominate the developing international nuclear power industry. Westinghouse and General Electric began to set up global commercial ties that would later yield billions of dollars in future royalties and export revenues.

TURNKEYS AND BANDWAGONS

At the beginning of the 1960s, nuclear power was still not quite competitive with coal or other conventional U.S. sources of electricity generation. The first straight commercial nuclear power plant purchase was finally made in December of 1963. In the first sale of what became known as the "turnkey era," G.E. sold a reactor, the Oyster Creek plant, to Jersey Central Power and Light.

The turnkeys were so named because G.E. and later Westinghouse, offered to design, supply, and construct the entire nuclear power plant at a low, fixed price. The utilities had only to "turn the key" and begin operating the plant. This extraordinary deviation from normal construction practice shifted most of the venture's risk onto the vendor. The gamble was that the vendor-subsidized prices would seduce utilities into ordering more nuclear reactors and would thereby foster a healthy market.

G.E. executive Bert Wolfe explained that "the turnkeys were the thing that turned nuclear power from a one-of-a-kind, uneconomic venture into an honest-to-God competitive product. They got enough volume in the business that you could build an engineering staff, standardize your product, and put up facilities to mass-produce things so that costs went down."

The costs did not go down far or fast enough to spare G.E. and Westinghouse a financial soaking estimated at close to a billion dollars between 1963 and 1967. But no other manufacturer sold a reactor until 1966. By then, the struggle for industry leadership involved only those two firms.

Whether planned or accidental, the turnkey years tre-

"TO ERR IS HUMAN, TO SQUEEZE OUT EVERY BUCK POSSIBLE IS DIVINE"

Copyright 1979 by Herblock in the Washington Post.

mendously boosted interest and confidence in U.S. nuclear power. Domestic orders jumped from seven in 1965 to 20 in 1966, and surged to a then record high of 30 in 1967.

These were aptly christened the "Great Bandwagon Years" by Phillip Sporn, then president of the American Electric Power Company. Despite a complete lack of figures on how much it cost to operate nuclear reactors (only eight were even operating by 1967), sales continued at a brisk pace into the 1970s. With the AEC and the reactor vendors confidently proclaiming the profitability of nuclear power, the utilities saw each additional sale as further proof of the commercial viability of light-water reactors.

The rush to nuclear power in the late 1960s caused the AEC to speed up its program to develop a new generation of nuclear reactors, the breeder reactor. The name "breeder" comes from the idea that this reactor can produce, or breed, more fuel than it consumes. Nuclear proponents argued that commercialization of the breeder was essential if the U.S. nuclear program was to avoid running out of uranium. The AEC's commitment to building a prototype breeder project at Clinch River, Tennessee, in 1972 was thus interpreted by the industry as a crucial symbol that nuclear power was here to stay in the United States.

The optimism was reinforced by the amazing commercial success of the early 1970s. Domestic sales averaged a walloping 30 orders a year then, and soared to a still unsurpassed high of 39 in 1973. There was talk during the Nixon administration of having 1000 nuclear power plants in use by the year 2000.

After yet another strong posting of 27 orders in 1974, the bottom fell out of the nuclear-reactor market. In response to the drop in the growth of electricity demand induced by the oil price hike of 1974, electric utilities deferred scores of plant orders. Since 1975 there have been barely a dozen new domestic orders, causing many to wonder how much longer the Big Four reactor vendors will stay in the business. *Nucleonics Week,* the major trade publication, wrote in November 1977 that "in the opinion of many, the giant U.S. nuclear industry is slowly, very slowly, bleeding to death."

General Electric is losing an estimated $30–40 million every year on its nuclear business. The losses for Combustion Engineering and the Babcock and Wilcox subsidiary of J. Ray McDermott, who have smaller investments in nuclear power, run about $10 million apiece annually. Of the Big Four, only Westinghouse claims to be making any nuclear profits. Even this is a dubious claim, since some utility companies have sued them for more than a billion dollars worth of contracted-for but undelivered uranium fuel. (Westinghouse contracted to get the uranium when its price was about $8 a pound. By the time the delivery dates arrived in the mid-1970s, the price had soared to over $40, and Westinghouse tried to get out of its obligation.)

The *de facto* moratorium on new reactor orders in the United States since 1975 has a lot to do with the industry's disappointing performance. But the companies were aware that profits were a long-time proposition when they first began committing major chunks of investment money to nuclear in the mid-1960s. A high-technology, capital-intensive business like supplying nuclear reactors and components entails massive startup costs. Hundreds of millions of dollars were spent to erect manufacturing facilities, to test products for performance and safety, and to hire the hundreds of engineers needed to design the plants. G.E. alone claims to have poured nearly $5 *billion* into its nuclear-energy business. The corporate strategy assumed that these costs would be recovered in later years, but it has not yet worked out that way.

"We have not yet reached the point of break-even on the nuclear business," admits Babcock and Wilcox Nuclear Vice President John McMillan. The same is true for Combustion Engineering and General Electric. Westinghouse, on the other hand, claims that its nuclear profit rate exceeds its overall corporate average.

Bert Wolfe of G.E. disputes the claim. "To my knowledge, nobody has been profitable in this business, including Westinghouse. You can play this game that they're profitable if you forget the uranium [litigation]. But the reason they sold so many plants is because they sold so much uranium."

PLAYING MONOPOLY

How can the nuclear companies afford to stay in a business that after 15 years has yet to turn a profit? Where will the profits come from?

The answers are varied and complex, but all lead back to one general theme that characterizes the structure of the entire nuclear industry—monopoly control by giant, diversified corporations.

Photo by Daniel S. Brody.

The laws of supply and demand described in economics textbooks don't apply in this world. Instead, corporate monopoly rules. Take prices, for example. Rather than being freely determined in the marketplace, prices tend to be "administered" to deliver a certain target rate of profit, regardless of demand. Or perhaps they are bid up to falsely extravagant levels, as the international cartel of uranium companies managed to do a few years ago.

The uranium sector graphically illustrates the concentration of economic power that typifies the nuclear industry as a whole. According to figures provided by the U.S. Senate Energy Committee, almost 50 percent of the nation's known uranium reserves are controlled by five companies. Leading the pack is the Kerr-McGee Company, with 21 percent of the market.

That Oklahoma-based mining company also holds the top spot in the second stage of the nuclear fuel cycle, by controlling one-fourth of the total capacity for milling uranium. Here, the top five firms control almost 70 percent of the market.

The heavy involvement of petroleum companies like Exxon and Gulf in the uranium business adds another twist to the monopoly game. The U.S. Senate study also revealed that 13 of the top 20 corporate holders of U.S. uranium reserves are oil and gas companies. They control over half the total reserves as well as over 40 percent of the uranium milling capacity. They are also heavily involved in the fabrication of fuel rods, another critical stage in the fuel cycle.

The oil companies' takeover of the uranium business is only one part of their grand corporate strategy to become *total* energy companies. The industrialized world is clearly entering a transition phase between the petroleum-based economies of the twentieth century and whatever mix of energy technologies fuels societies after the year 2000. Oil companies like Exxon want to maintain their dominant status, so they are making sure they prepare for all alternatives.

To that end, they are using the mountains of cash generated by their monopoly control over world oil supplies to buy the same monopoly control over *all* energy sources that look promising. They are gobbling up the coal business and trying to wrap their fingers around solar energy and other alternative technologies.

Seeing that nuclear power had the potential to become perhaps *the* energy source of the twenty-first century, the oil companies decided to get into the business. Although some of them had been involved in uranium from the time the government first nurtured the industry into existence back in the 1950s, their present monopoly position was secured mainly through acquisitions funded by the 1973 oil-price explosion.

Their decision to go after the uranium sector of the nuclear business was a brillant strategy. Rather than spend billions of dollars to muscle into the reactor market before nuclear power was a sure thing, Exxon and other oil companies instead chose to spend far less by controlling the fuel for nuclear power. If nuclear does finally take off, they will still have their hands on the fuel faucet.

The uranium business cannot compare, however, with the monopoly to be found in the nuclear-reactor business. The massive sums of investment required meant that only the largest and most experienced firms could hope to succeed at it. The Big Four vendors fit the bill exactly. Each has been supplying equipment to electric utilities for decades, and their combined corporate assets total $25 billion.

The four supply virtually all major pieces of heavy equipment that make a nuclear power plant run—the nuclear steam supply system, or reactor; the steam generators and pressure vessels; and the heavy turbines. Westinghouse and G.E. have traditionally split about 70 percent of the domestic nuclear reactor market, with Combustion and Babcock sharing the remaining 30 percent. This is not to mention the fact that G.E. and Westinghouse reigned almost absolutely over the international market until the early 1970s. Along with Rockwell International, a multibillion-dollar company that has long been one of the Pentagon's favorite contractors, the four are also the core companies working on the government's breeder development program.

The four control, besides equipment, the supply of ready-to-use reactor fuel. Exxon, with far less than 10 percent of this market, is their only serious competitor. The highly profitable service, repair, and spare parts businesses are also controlled by the vendors.

THE DROUGHT BEGINS

In the early 1970s, it looked as if the vendors would be making the nuclear superprofits that monopoly usually delivers. But the disastrous slump that began in 1975 has forced the companies to reevaluate their plans.

Behind the slump in new reactor orders, now in its fifth year, lies a harsh world of energy reality. The OPEC price explosion in 1974 raised energy costs to the point where broad-based conservation efforts made economic sense. The price rise induced a drop in America's demand for electricity, which had galloped along at a seven percent annual growth rate for years, until 1975. It still remains well below the traditional rate. Utilities, caught with excess capacity for generating electricity, had little choice but to defer existing plant orders and delay new ones.

The economics of nuclear power, originally its major selling point, also began to fade. The double-digit inflation that hit the United States in 1974 made borrowing the huge sums of money needed to finance nuclear power plants extremely difficult. Capital costs for nuclear power continued to climb as well. The price of a 900-megawatt plant increased fourfold between 1972 and 1977, as a result of safety-related government-mandated reactor design changes, management's difficulty with construction schedules, and—to a lesser degree—environmentalist opposition. It now takes 10 to 12 years to bring a nuclear power plant on line.

The industry was also losing ground on the political front. Nuclear opponents scored a major victory when they forced the dissolution of the AEC in 1974 and the JCAE two years later. The Nuclear Regulatory Commission (NRC) was created to take over the AEC's job of regulating nuclear power, while the Energy Research and Development Administration, a predecessor to today's Department of Energy, took over nuclear promotion and weapons production.

In the Congress, nuclear oversight was spread across several different committees to try to preclude the institutionalized bias toward nuclear power that had formed in the JCAE over the years. The appropriations committees in both houses, which must approve the budgets for nuclear programs, remain critical pressure points, as do the commerce committees, the Senate Energy Committee, the House Interior Committee, and the House Science and Technology Committee which also exercise great control over nuclear matters.

The election of Jimmy Carter to the presidency in 1976 further complicated nuclear power politics. Carter was elected partly through the efforts of environmentalists, who were attracted by his apparent opposition to nuclear power. Since his inauguration, Carter has pleased nobody on either side of the issue. Environmentalists feel betrayed by Carter's selection of nuclear hard-liner James Schlesinger to head the Energy Department, the increased nuclear dependence Carter's energy plan calls for, and the President's continued financial support for breeder and other "hard" technologies instead of solar power.

From the industry standpoint, the Carter administration has been an absolute disaster. Its sins include: failing to expedite the reactor licensing process; delaying a solution to nuclear power's long-term radioactive waste-disposal problem; handcuffing U.S. reactor exports because of what industry sees as misplaced concerns about nuclear weapons proliferation; jeopardizing national security by trying to cancel the Clinch River Breeder Reactor project; and the President's failure to publicly retract his campaign promise to use nuclear energy "only as a last resort."

Then, in March 1979, just when the ailing nuclear industry could least afford it, the almost catastrophic accident at Three Mile Island dealt another serious blow to the nuclear cause. The more than 100,000 protestors gathered in front of the Capitol on May 6, and the strong media interest, indicate that nuclear power will be a major issue in the 1980 presidential election. Things look bad enough that even such a bastion of corporate opinion as *Business Week* magazine has wondered whether nuclear power is on its last legs as a profitable business proposition.

DOWN, NOT OUT

For its part, the nuclear industry is by no means ready to call it quits. The industry's post-Harrisburg statements indicate it still believes there are profits to be made in nuclear power. Even the J. Ray McDermott Company plans to stick it out. C. Duvall Holt, public information officer at its Babcock and Wilcox subsidiary (builder of the ill-fated Three Mile Island plant), dismissed speculation that the company would forsake the business.

"Heck, no," he said. "We're learning a lot from this incident, and we see quite a healthy business in the years ahead." Representatives from Combustion Engineering and G.E. voiced similar sentiments.

If the companies lost faith in the likelihood of nuclear profits, they would doubtless leave the business. That has not yet happened. G.E.'s Wolfe said, "We believe there are still profits to be made in the nuclear business, especially in the area of refueling and servicing existing reactors."

The decision to stay is not predicated mainly on a desire to recoup past losses. As Wolfe says, "That's all past history. It's questionable whether we'll ever be cumulatively profitable, at least in terms of discounted rate of return." The point is that past losses have already been fully digested by the corporations. Their decision to stay in the business now depends mainly on expectations of what the future will bring. And since each of the Big Four has already made the basic investment in nuclear production, the "extra cost" of remaining poised for the market's presumed return is insignificant compared to the potential profits.

Getting out of the business is not as easy as it sounds, either. Because of nuclear plants' long lead times, the companies are contracted to deliver reactors up through the early 1990s. This "backlog" of nuclear work, which totals over $10 billion for Westinghouse and G.E. alone, helps the companies keep their teams of nuclear engineers and workers together while waiting for the market to return. So even if the Congress declared a moratorium on new plants tomorrow, the nuclear business would continue for at least another decade.

In fact, right now nuclear executives are extremely hopeful about profits to be had from existing plants. The profit rate for refueling and servicing today's reactors is about 1½ times the 10 to 12 percent pre-tax margin that vendors get on reactors themselves. And the market promises to expand as more of the plants now under construction finally come on line. Combustion Engineering, for instance, expects a business of about $200 million a year in refueling and servicing on the 31 plants it will have on line by 1992. Conservatively estimated, that translates into $30 million (in 1979 dollars) of nuclear profit, even if Combustion never gets another reactor order.

Although the companies have no prejudice against such "old" profits, they desperately would like to see the reactor market return. They expect this to happen, although not for another year or two. The electrical utilities are still stuck with excess capacity. But the industry figures this will work itself out in a couple of years and that orders will then begin to average about 12 to 15 reactors a year. Here is where companies like Westinghouse and especially G.E. can use their bulk to great advantage.

Each is a giant, diversified multinational corporation, by no means dependent on the immediate return of the nuclear market to ensure overall corporate health. G.E.'s corporate assets are $15 billion. Westinghouse's are $6 billion. The other two vendors weigh in at about $3 billion apiece. Nuclear is only a small fraction of the total corporate business for these firms. It accounts for only 3 percent of G.E.'s total sales, and about 10 percent of total corporate sales for the other three members of the Big Four. Westinghouse and General Electric, in particular, can afford to "wait" for the market's revival.

Westinghouse, in fact, plans to do a good deal more than that. Unlike the other vendors, who have put a hold on new nuclear investment, Westinghouse just made the second largest investment in its history, spending $50 million to construct a zirconium plant. "Over the next five years," Westinghouse Power Systems President Gordon Hurlbert modestly commented, "a couple of hundred of million is all I expect to invest in the nuclear business."

The nuclear industry believes that the energy mess will worsen in the 1980s, leaving the United States with no alternative but nuclear power. Hurlbert expects U.S. suppliers to get about 20 orders a year from the domestic and the global markets through 1982. He bases his optimism on a cold assessment of the international energy situation. In a January interview he said, "Just watch—if they don't get the Iran situation straightened out by June, there's going to be gas lines to where you can't stand it. And there will be massive new orders for nuclear power plants. A year, two years, three, it really doesn't matter when it comes. I'm looking at nuclear power to pay my pension when I retire.

PLAYING BALL

Other sectors of the business community share Hurlbert's colorful assessment of the nuclear imperative. In an April editorial *Business Week* warned about the dangers of *not* going nuclear. "The U.S. will remain at the mercy of the oil-producing cartel until it develops an effective energy program, which must include increasing nuclear capacity. . . . The rest of the world obviously does not intend to stand still, regardless of what this country does."

Nor has Wall Street's enthusiasm for nuclear waned after the Harrisburg accident. Investment banker Kemp Fuller, vice president at Moseley, Hallgarten, and Estabrook, believes that "the nuclear industry is not going to go away under any circumstance [and] people are going to have to learn to live with it."

In order to attract the large sums of fresh investment needed to fuel the future growth of the nuclear business, investors must be sufficiently impressed with the business's growth potential. Hence, the attitude of commercial banks, insurance companies, investment banks (who essentially act as stock brokerages), and the other financial behemoths that manage the economy's investment flow is critical to the future of nuclear power.

Compounding the indirect influence that the Big Four's major institutional investors like Morgan Stanley and Chase Manhattan exert over the vendors' nuclear strategies, Wall Street also exercises a strong, direct power over nuclear's future through its tight relationship with the electrical utilities. The utilities depend on Wall Street to lend them the large sums of money needed to finance power-plant construction. In the past, utilities were known as "widow's stock," since they consistently rewarded their investors with attractive, albeit unspectacular, dividends.

Wall Street and the utilities have traditionally favored nuclear investments for similar reasons: nuclear plants' greater capital costs translate into higher profits, while simultaneously solidifying centralized control over electrical power production. After Harrisburg, "the financial community will still play ball on nuclear with utilities," says Allan Benasuli, an analyst at the investment bank of Drexel-Burnham-Lambert, "but on tighter terms than before." The Virginia Electric Power Company (VEPCO), the first heavily nuclear utility to approach Wall Street for a loan after the accident, found Benasuli's comments to be right on target. VEPCO attracted investors but only by accepting a significantly higher interest rate than usual.

In weighing possible future nuclear investments, Wall Street will take into account how the Pennsylvania Public Utilities Commission handled the costs of the Three Mile Island accident. The Commission decided in June, 1979 to pass along about 80 percent of the estimated $15 million monthly cost for buying replacement electricity to ratepayers. Had the entire charge instead been levied against the stock and bond holders of the General Public Utilities Corporation, who own the TMI plant, investor confidence in nuclear power might well have been destroyed.

GPU Chairman William G. Kuhns had previously told a Senate subcommittee that unless the utilities were permitted to charge their customers for accidents like TMI, Wall Street would jack up financing charges to utilities by over 20 percent. The increase, Kuhns said, would ultimately contribute to more expensive electricity, "so consumers wind up paying one way or another."

Kuhns and other industry representatives have suggested that the costs of any future nuclear accidents be spread among the nation's consumers, in order to maintain investor confidence. In that case, utilities could conceivably begin to order nuclear plants again—once they work off their current over-capacity.

A FEDERAL ROLE

Finally, the future health of the U.S. nuclear industry depends largely on its original creator and protector—the federal government. The industry was weaned on government subsidies and remains addicted to them today. This raises a major contradiction. The industry could not survive without Washington's support in the form of research and development, insurance subsidies, and the like. Yet the sacrifice of autonomy and the vulnerability to the whims of Washington that are entailed necessarily make the business more than slightly unpredictable. The primary attraction of monopoly control—the reduction of risk —is confounded by the industry's dependence on the state.

To be sure, the government could do plenty to turn the industry's problems around. The Carter administration could, for example, relent to heavy pressure from the industry to reprocess the spent fuel from nuclear reactors.

In reprocessing, the portion of spent fuel that can be reused is extracted so that it can be used to fuel other reactors. Nuclear opponents fear that the plutonium produced during reprocessing will eventually seep into the atmosphere and cause incalculable damage to the public's health—or fall into the hands of terrorists and nations seeking their first atomic weapons.

The industry claims that reprocessing is safe and that it will multiply the amount of uranium available for nuclear fission by a factor of 70, thereby offering a long-term solution to the energy crisis. The guarantee of ample fuel would also rekindle utility interest in ordering new nuclear plants.

With so many obvious advantages, it may seem surprising that the industry itself is not operating reprocessing plants. It's not that they haven't tried. It's just that the private sector's experience with reprocessing facilities has been an unmitigated financial disaster.

Nuclear Fuel Services, a subsidiary of Getty Oil, gave up on its West Valley, New York, plant in 1972 after six years of unprofitable operation. It left the State of New York with the roughly $1 billion job of disposing of more than 600,000 gallons of radioactive waste products. General Electric pulled out of its $64 million Morris, Illinois, reprocessing venture in 1975. G.E. doubted that the plant could ever meet government safety regulations and still turn a profit no matter how many millions were poured into it.

The last private reprocessing plant, a joint venture of Allied Chemical, Gulf Oil, and Royal Dutch Shell, operating under the name Allied General Nuclear services at Barnwell, South Carolina, was scheduled to open last year. But President Carter's April 1977 statement "indefinitely deferring" reprocessing in the United States put a crimp in the plans. In response to the NRC's refusal to license the plant, the companies have sued the federal government to obtain the money needed to complete the plant. A presidential decision to allow the Barnwell plant to begin operation would remove an important obstacle to future nuclear power development in the United States.

The federal government could also work harder to help G.E. and Westinghouse regain their once unquestioned command of the international market for nuclear equipment.

G.E. and Westinghouse have traditionally butted heads in many different markets. When they opened the domestic nuclear market via the turnkey strategy in 1963, the struggle intensified. Nuclear power was heralded as one of the major new growth areas for American big business, and the two companies wrestled hard for the industry's top spot. The winner figured to reap the benefits of greater sales and correspondingly higher profits, as well as intangible prestige advantages.

The two multinational giants shared the world market during the 1960s, amassing such a monopoly that by 1972 over 90 percent of the world's reactor exports bore the Americans' trademarks. Each staked out an impressive empire of affiliated foreign companies in Japan, France, Spain, Italy, West Germany, and elsewhere.

G.E.'s Bert Wolfe explains the goal of the corporations' international strategy: "In the 1960s when we were planning our export markets, we realized that our markets would be limited by the fact that the Europeans would develop their own nuclear capability. That was partly why we chose to push for licensing agreements, rather than saturate them with reactor orders."

By selling, or licensing, its nuclear knowhow to European and Japanese firms, the two American firms assured themselves of more than the spectacular short-term profits that came from actually selling reactors. The licensing fees and the royalty payments that France's Framatome company pays to Westinghouse have been, according to a knowledgeable Wall Street source, a major share of the $20 to $30 million in profits annually generated by Westinghouse's nuclear business.

The licensing strategy ultimately backfired. The European and Japanese firms quickly assimilated the U.S. light-water technology. They got plenty of government help, just like their American counterparts. Nuclear research and development were subsidized and trade walls were erected to keep the Americans out.

A study by Richard Barber and Associates aptly describes

the U.S. companies' situation, saying, "[The] industrialized countries . . . have now protected their infant nuclear reactor industries from direct outside competition in their own markets, thus forcing U.S. producers to seek new export orders in Third World countries."

The two American multinationals did exactly that with great success, selling 17 reactors to the Third World between 1971 and 1974. Such sales are especially lucrative because there is usually no competition from entrenched domestic companies in a country like the Philippines. The exporter is thus able to provide a larger share of the overall product and may also widen its profit margin.

American domination of this market, however, has also been overthrown. For one thing, Third World sales have been harder to come by since the 1974–75 world recession slashed poor countries' export revenues and worsened their already precarious burdens of debt. Most recent orders have come from Brazil, Pakistan, the Shah's (prerevolution) Iran, and other rising regional powers interested in obtaining nuclear technology's badge of stature. U.S. companies faced tough foreign competition for these contracts, as Framatome and West Germany's Kraftwerk Union company schemed to break the American stranglehold.

The American firms finally lost much of this business. U.S. government regulations did not allow the export of uranium-enrichment or reprocessing facilities, for fear that the recipient country, particularly ambitious nations like Brazil and Iran, would use them to fashion nuclear weapons. Unencumbered by such restrictions, Framatome signed a deal with Pakistan, and Kraftwerk Union landed a multibillion-dollar contract in 1975 to provide Brazil with a complete nuclear fuel cycle.

The American firms' comparative disadvantage was later institutionalized when Congress passed the Non-Proliferation Act in March of 1977. As the domestic market continued to lag, U.S. nuclear executives grew increasingly adamant in their demands for a more favorable government export policy.

The Carter administration responded late last year with promises of increased Export-Import Bank funding for nuclear exports and quicker bureaucratic approval of export applications. If this policy shift is reinforced, with others like it, U.S. reactor exports could swell to their former levels and save the industry from an early grave.

Jimmy Carter has made other attempts to befriend the nuclear industry. The most important came in a clandestine White House meeting held June 14, 1978, with the chiefs of the industry's 13 biggest companies.

Carter had been trying to cancel the Clinch River Breeder Reactor Project since 1977 because he feared that the commercialization of the plutonium-producing machine would intensify the global proliferation of nuclear weapons. The nuclear industry had relied on its greater strength in Congress to deny the President's plan. At the unpublicized White House meeting, the Carter administration offered a compromise. If the industry would give in on Clinch River, the President promised to design an even bigger, better breeder that could be built after the 1980 elections. Even more important, though, was the administration's offer of something that nuclear executives say they need more than anything else to combat the public skepticism that is crippling the nuclear cause. The President would issue a strong public statement endorsing the safety and necessity of nuclear power for the United States.

The industry declined Carter's offer rather than give up on Clinch River. Their refusal gives clear evidence that they are hardly ready to give up the ship.

Why did the nuclear power industry refuse a deal that might have guaranteed a nuclear future for the United States? They believe, rightly or wrongly, that such a future is already inevitable. Gordon Hurlbert, the head of nuclear power at Westinghouse, echoed sentiments of many in the

industry when he said, "There's really a nuclear imperative. If we're going to raise the standard of living for the
world, really there isn't any choice but the light water
reactor . . . and ultimately the breeder reactor." Why give
up the Clinch River program for a statement that would
only hasten the inevitable?

INDUSTRY DIGS IN

We could debate whether the corporations are right
about the inevitability of nuclear power for the United
States. But in some ways this is of secondary importance.
What is most critical is that the firms believe it and plan a
no-holds-barred political campaign to guarantee it. The
national debate over nuclear power engendered by the
Harrisburg accident threatens tens of billions of dollars in
past investments and future profits. With that kind of
money on the line, the industry has no choice but to push
even harder for a nuclear economy.

Nuclear opponents should beware of premature elation.
The nuclear industry still has plenty of ammunition in reserve. Taking a page out of the energy blackmail book
written by their cohorts in the oil business, they will claim
that unless we accept nuclear power, we will become
slaves to OPEC, and suffer higher energy prices, higher
unemployment, and inflation.

There is another alternative. Nuclear power currently
provides less than 4 percent of America's total energy. A
recent report by the President's own Council on Environmental Quality concluded that energy savings of up to 40
percent—10 times nuclear power's contribution—are possible right now, with no reduction in human comfort. The
savings would be realized by substituting technologies that
are available right now, such as home insulation, more
efficient appliances, and improved automobile designs.

Such a strategy directly conflicts with the wishes of the

nuclear industry. They have resisted it and will continue to resist it. Only time will tell whether the corporate view will prevail. The challenge for nuclear opponents is to mobilize the American public to demand and work for an energy policy that serves the best interests of a majority of Americans—not one that exposes us to the risks of nuclear power while picking our pockets so that a few giant corporations can reap higher profits.

Doing Without
Nuclear Power

Charles Komanoff

The case for nuclear power has always rested on two claims: that reactors were reasonably safe and that they were indispensable as a source of energy. Now the accident at the Three Mile Island nuclear plant in Pennsylvania has shaken the first claim, and we will soon have to face the flaws in the second. The result should be the abandonment of nuclear power and the emergence of a more rational energy policy, based on measures to improve the efficiency with which energy from fossil fuels is used.

The dangers of nuclear power have been greatly underestimated, while its potential to replace oil as the world's primary energy source has been vastly exaggerated. Rather than being indispensable, nuclear power can make, at best, only a modest and easily replaceable contribution to future energy requirements.

Nuclear power plants currently generate 13 percent of the electricity produced in the United States, and slightly smaller percentages in Western Europe and Japan. (Countries such as Sweden and Switzerland, which depend on nuclear reactors for 20 percent of their electricity, are exceptions.) Since electricity accounts for only 30 percent of the total energy supply, nuclear power provides less than 4 percent of the overall energy of the industrial countries. More surprisingly, as the economist Vince Taylor has shown in a report to the U.S. Arms Control and Disarmament Agency, *Energy: The Easy Path,* nuclear power could provide, at most, only a 10 to 15 percent share of the energy supply of the advanced countries by the year 2000.

Taylor's argument begins with the fact that nuclear power provides only electricity, an expensive form of energy which absorbs only 10 percent of the oil used in the United States and other advanced countries. The great hopes for nuclear energy were based on the possibility that electricity could be substituted for oil in processes where electric power had not been heavily used. In fact, little such substitution has occurred. The petrochemical and transport industries, including automobiles, now use 50 percent of available oil. There is no foreseeable technical possibility of electrifying large proportions of these industries. The remaining 40 percent of oil is used for space heating and for industrial energy. In both cases, major electrification is ruled out by nuclear power's high cost.

What are the relative costs of energy from nuclear power? They are rarely compared to the costs of burning oil in engines or furnaces, yet such a comparison is central if we are to weigh the prospects of substituting nuclear power for oil. If we carefully estimate the different kinds of costs involved in producing kilowatts at nuclear plants and in obtaining barrels of oil, we find that the cost of energy from a nuclear plant built today can be calculated at $100 for the "heat equivalent of a barrel of oil." This figure reflects fuel, maintenance, distribution, and—most important—the cost of constructing the nuclear plant itself. It is four times the cost of heat from oil at OPEC prices.

The heat in electricity can be made use of with more efficiency than can heat from oil and gas—as much as twice as efficiently. For example, industrial electric furnaces produce about 50 percent more glass, per unit of applied heat, than gas-burning furnaces. When we allow for this, electricity from new nuclear plants still remains two to three times as expensive as oil. As a result, the market for nuclear electricity has slowly but inexorably been drying up. Notwithstanding plans to build hundreds of reactors, the projected demand for nuclear power—and with it the financial backing for its expansion—has not materialized.

It wasn't always this way. At the start of 1972 the first plants purchased by the electric utilities on a commercial basis were entering service at a cost of $200 million for a standard 1000-megawatt generator. This cost converts to the equivalent of about $25 per barrel of oil. Costs were expected to fall as more plants were built. The tripling of prices for crude oil and coal in the wake of the 1973 Arab oil embargo was expected to put nuclear power in a commanding market position. Not only was it apparently cheaper than coal as a source of electricity, but nuclear

"OH, MY GOD—IT'S A MELTDOWN!"

electricity cost only twice as much per British thermal unit
—i.e., per unit of delivered heat; and since this heat was
available in more efficient form, it was therefore close to oil
in energy value. Moreover, nuclear electricity was expected to decline in cost while oil rose. Few doubted the
Atomic Energy Commission's projection in 1974 that one
thousand 1000-megawatt reactors would be built by the
end of the century—capable of producing roughly the
equivalent of the entire U.S. energy supply in 1972.

Instead, nuclear costs have soared as the industry expanded, and primarily for two reasons: safety systems had
to be added to meet rising public concerns; and improvements in design were needed to correct defects that
emerged as operating experience increased. By the end of
1977, a 1000-megawatt reactor cost $800 million, four
times the cost of a 1972 plant. The rate of increase was an
astounding 26 percent per year. This was triple the rate of
general U.S. inflation and half again as great as the increase
in construction costs of power generators using coal even
if the coal plants have the best pollution controls. It was
about the same rate of increase as that for the price of oil,
induced by the actions of the OPEC cartel. Nuclear power
had thus lost its edge over coal in the electric power market and had lost the opportunity to make inroads on the
larger market occupied by oil and gas.

To be sure, these figures represent the findings of the
author's own cost research. They have not as yet been
digested by the power industry, which still touts nuclear
reactors as the least costly source of electricity. The Edison
Electric Institute claims that power from nuclear reactors
costs 1.5 cents per kilowatt-hour, as compared to 2 cents
for power from coal and 3.9 cents for power from oil. But
this claim is based on a sample dominated by nuclear plants
completed before expensive safety standards were imposed and including coal and oil plants operating at low
efficiencies because of the surplus of generating capacity.

More important, this comparison pertains only to the 10 percent of oil which is used to generate electricity.

Nevertheless, the impact of steeply rising costs and concomitant delays in licensing and constructing reactors has not been lost on nuclear advocates. Before the breakdown at Harrisburg, they were stridently demanding that Congress and the regulatory authorities "stabilize" reactor design standards as a solution to rising costs. Now the Harrisburg accident has raised the prospect of a reactor disaster from an infinitesimally remote possibility to a reality. This guarantees that safety requirements will be stiffened so that costs will continue to rise sharply, offsetting probable increases in the cost of oil. The events at Harrisburg have pushed nuclear power beyond the brink of economic acceptability.

Moreover, the accident has undermined confidence in the nuclear industry. As the laxity of safety regulations at Three Mile Island becomes widely known, it may undermine confidence in the Nuclear Regulatory Commission as well. The commission had composed, in its 1975 Rasmussen Report, a supposedly exhaustive list of possible "initiating chains" for nuclear accidents and concluded that, in view of its precautions, U.S. reactors were statistically safe. The sequence of events which caused the Harrisburg accident was not among these "chains."

Henceforth every major decision about nuclear power, especially those concerned with disposal of radioactive wastes, will require so much public scrutiny that delays and costs will become intolerable. Well before the Harrisburg events, seven state legislatures had already passed laws to stop, or control, the storage and transport of nuclear wastes. We can now expect that pressure for more stringent legislation will mount.

Of course, for many people the prospect of permanent oil shortages causing economic stagnation is worse than unsafe and expensive nuclear power. And, in fact, nuclear power will be justified in the near future by references to the world's limited store of easily extractable oil. The median estimate of the total quantity of world oil that remains to be exploited is 1½ to 2 trillion barrels. This would be enough for a 100-year supply at the current rate of consumption. It will not suffice if world needs grow at several percent per year—the rate of energy growth generally considered necessary to support healthy economic growth. At 3 percent annual energy growth, for example, this supply would be exhausted in 50 years. And well before then, physical limits on the potential rate of discovery and extraction would force the level of oil output below demand.

The possibility that potential oil resources may be greater than is now anticipated does not significantly alter this pessimistic picture. Aside from the doubtful wisdom of tying the world's economic prospects to unexpected discoveries, even a doubling of total oil resources would add only another 20 years of supply, assuming 3 percent annual growth in consumption. Oil is simply incapable of sustaining world energy growth for more than a few decades.

Nor are coal or natural gas, the other fossil fuels, likely to take up much of the slack. Natural gas is less abundant than oil. Moreover, continued growth of gas consumption would require that it be transported in liquefied form—a matter of growing controversy—and would maintain the present undesirable level of dependence on imports from the Middle East.

The world's recoverable coal reserves are several times those of oil and will now be increasingly exploited in the United States. But environmental and cost constraints analogous to those impeding nuclear power will probably prohibit developing coal on the scale necessary to increase energy growth significantly. During the past decade, efforts to reduce injuries and deaths in U.S. mines and to contain pollution from coal burning have caused the cost of coal-fired electricity to increase at twice the general inflation rate. Any increase in the rate of expansion of the coal industry would probably accelerate this trend.

Moreover, hopes of manufacturing large quantities of synthetic oil and gas from coal or shale appear doomed by inherent inefficiencies and environmental effects. The costs of dealing with both would be staggering. According to a study by energy specialists at the Massachusetts Institute of Technology under the direction of Carroll L. Wilson, the production of synthetic fuels equivalent to only one-tenth of the current U.S. energy supply would require mining the equivalent of the entire U.S. coal output in 1975. To process this fuel would require 42 gasification facilities and six oil shale plants, costing over $1 billion each, as well as 47 liquefaction facilities, each costing one-half billion dollars. Oil and gas from such facilities would be several times as costly as OPEC oil and gas.

Furthermore, since no commercial-size synthetic fuel plants are operating as yet, little attention has been devoted to the environmental effects of these technologies. If we follow the known pattern for energy projects with potentially large effects on the environment, we can estimate that the construction costs currently cited for commercial synthetic fuel plants could easily double or triple, ensuring that such fuels would remain uneconomical.

Similar considerations apply to nuclear fusion and solar space satellites. Even if the considerable scientific and engineering problems can be solved, high costs will consign these "exotic" high-technology schemes to a minuscule share of energy supply.

Aside from nuclear power, only solar energy—which includes, in addition to sunlight, other sun-driven sources such as wind, water, and plants—appears to hold the technical potential to provide sufficient energy to support the world's expanding economies. However, it will take a long time for solar energy to become the dominant source of world energy. It will first be necessary to eliminate inefficiencies in solar technologies, to allow turnover of much of the existing housing stock and industrial machinery, which is unsuited for solar energy, and to permit national economies to absorb solar energy's high costs. Solar technologies are now generally less expensive than electricity from new nuclear plants but cost considerably more than oil.

The earliest anticipated date by which this transition could be accomplished in most advanced countries is 2025. But before then, it is likely that oil extraction could not keep up with increasing oil requirements, assuming that the amount of energy in use increases by several percent a year. Some experts argue that nuclear power, despite its high risks and poor economic prospects, is therefore needed at least as a supplementary, transitional energy source, to avert major shortages of fossil fuels, which would imperil economic growth.

However, this view ignores perhaps the most significant —and certainly the most neglected—factor in the current discussion: the large potential for reducing the amounts of energy used, such as oil or gas, without affecting the quality or quantity of energy services, such as heating, lighting, and transportation. Energy must be combined with other materials, equipment, and labor to provide energy services. The amount of energy required to provide a given service can thus vary widely, depending upon the amounts of other resources used and the technology employed. As Vince Taylor points out in the paper referred to earlier, the *productivity of energy* is not fixed but is susceptible to deliberate change.

Technological innovations, such as those occurring in solid-state electronics, will surely be one source of improvement in energy productivity, but major technical advances are not required for a successful program of productivity increases. A variety of simple measures to eliminate unproductively used energy and to improve process efficiencies could allow energy services to be expanded without increasing the amount of energy consumed.

For example, if the walls and ceilings of new homes were equipped with insulation, using only amounts which would pay back the added cost through fuel savings within 10 years, this would reduce fuel consumption by one-third or more, by comparison to consumption in typical houses today. A further reduction by one-half could be accomplished by a variety of building improvements: thick walls to moderate temperature changes; large, south-facing windows to capture winter sunlight; sophisticated thermal controls to match heating output to temperature desired; and better designed and maintained furnaces. These changes, along with insulation, could reduce fuel requirements for home heating to less than one-third of the present average, with no loss of comfort. Even this estimate does not exhaust the possibilities of reducing the energy consumed, since the addition of sufficient solar collectors and of heat-storage capacity could altogether eliminate fuel consumption for heating new houses.

The large potential for improving the productivity of energy extends to every sector of usage and is not confined to the United States. It has been demonstrated in detailed studies in virtually every other industrial country—even those thought to be much more efficient in their use of energy than the United States. For example, a recent analysis of the United Kingdom by the International Institute for Environment and Development concluded that applying currently available technologies could reduce energy requirements from today's level by 8 percent by the year 2025, while economic output would triple. Savings would come mostly from reducing heat losses in buildings, raising auto mileage through lightening and redesigning vehicles and the electronic control of engine operation, increasing the efficiency of industrial electric motors through proper sizing and coupling, and improving the efficiency of household appliances.

Broadly similar results have also been obtained in studies in France—where lack of indigenous fuels has been used to justify a huge commitment to nuclear power—as well as Denmark, Sweden, the Netherlands, West Germany, and Switzerland. The common conclusion of these studies is that cost-effective improvements in energy productivity can extend the lifetime of oil and gas resources sufficiently to obtain the time needed by the industrial countries for an orderly transition to solar energy.

Energy productivity has in fact been rising since the oil price increase of 1973. American factories have reduced their energy consumption per unit of output by 18 percent, resulting in energy savings in 1978 equivalent to three times the entire output of U.S. nuclear power plants.* In the United States as a whole, economic growth has recently been running at several times energy growth. Gains have been smaller in Europe, in part because of the lower rate of increase in the cost of imported oil (owing to appreciation in the value of most European currencies against the dollar), and also because of greater investment in electrification, which generally yields less economic output per unit of energy. Further improvements in energy productivity will come as energy prices rise, but a strategy aimed at eliminating so-called "institutional traps" inhibiting conservation could coax forth greater energy savings at less cost to consumers and with less impact upon the poor.

One such trap is the rate structure under which utilities sell electricity and gas. Rates are set by state authorities here and by national bodies in Europe. Generally they decline with increasing consumption so that reductions in usage resulting from conservation bring about less-than-

*This comparison assumes that nuclear's energy contribution is 1.7 times its "delivered heat" content, based on the greater efficiency with which heat from electricity can be used, compared to heat from burning fuels directly.

proportional reductions in electricity and gas bills. Such
rate structures made economic sense in the days when
expansion of supply led to more efficient production, which
in turn reduced average costs. However such rate structures make no sense today, when new, "incremental" supplies of electricity and gas are now available only at rising
costs. The effect of these structures is to discourage conservation, since, for example, a 20 percent reduction in electricity usage typically reduces customer bills by only 10 to
15 percent.

A second trap is the tax code. Virtually all industries that
supply energy in the United States benefit from credit and
other advantages accruing to businesses that are "capital
intensive." Utilities also enjoy special tax provisions obtained in recent years by pleading hardship from rising
costs for construction and fuel. Accordingly, new investments by electric and gas companies actually reduce their
taxes. According to Cornell University economist Duane
Chapman, deductions for interest expense, accelerated depreciation, and the investment tax credit are sufficient to
offset entirely the nominal 48 percent tax rate on corporate
income. By contrast, investments to improve energy productivity have far fewer advantages. Thus productivity-
raising ventures are frequently less profitable, even though
they are superior to supply investments in the energy obtained per dollar invested. This causes capital to flow toward expanding energy supply.

Finally, there are jurisdictional traps. Few houses are
constructed to be efficient in their use of energy, since
most builders seek to minimize equipment costs to keep
the original selling prices down. Similarly, much electricity
usage in commercial and apartment buildings is master-
metered, rather than individually billed, eliminating any
economic incentive to conserve. In industry, many potential energy-saving measures go begging, despite prospec-
tive rates of return as high as 50 percent; management

instead assigns a higher importance to conventional investments needed to remain competitive, such as expanding
plants and developing products. Yet regulated utilities can
attract capital, at lower rates of return, to expand their
generating capacity.

The result of these and many other barriers to productivity improvements is that individuals and businesses now
undertake only a small fraction of the available measures
that could improve energy productivity at a cost less than
the cost to the economy of supplying equivalent energy.

Most of these barriers have staunch defenders: large
power users who benefit from quantity discounts; energy
corporations whose profits depend on favorable tax treatment; home builders and appliance manufacturers who
contend that the cost of energy-saving measures will reduce sales. For example, lobbying by refrigerator manufacturers caused California's new energy use standard to be
set near the *lower* end of current refrigerator efficiencies.
However, the staying power of these institutional arrangements appears to result less from the political strength of
their beneficiaries than from the lack of a constituency for
reform.

The near disaster at the Three Mile Island nuclear plant
provides an opportunity for new departures in policy,
which would have been required for oil conservation in
any event, in view of the limits on nuclear power I have
described above. Because of short-term considerations of
cost and energy, most of the 70 nuclear plants now operating in the United States are continuing to run; another 50
where major construction has started may be completed.
But if improvements in energy productivity were made on
a wide scale, this could ensure that no additional plants
need be built. A similar outcome is possible in Western
Europe and Japan, where costs have also spiraled as the
result of the necessary but perhaps futile efforts to resolve
rising doubts over nuclear power's safety.

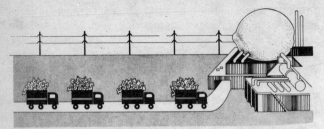

Credit: Environmental Action Foundation.

Nuclear power would then be left providing just over 5 percent of the industrial countries' energy supply, assuming overall energy requirements remain at today's levels. It would then be realistic to ask whether the dangers in plant safety, in accumulation of radioactive wastes, and in potential contribution to nuclear weapons proliferation do not exceed the minor benefits of continuing to run existing plants. To judge by the response to the Three Mile Island accident, that debate has already begun.

The Bargain Consumers Can't Afford

Richard E. Morgan

The year was 1966. Officials of the Consumers Power Company had completed plans for their Palisades nuclear plant and construction was under way. Within a few years, they promised, the $93 million facility would provide low-cost nuclear energy to the residents of central and western Michigan.

But Consumers' executives were in for a shock. By the time the Palisades plant was completed in 1971, it had cost $188 million—more than twice the original estimate. That was only the beginning of Palisades' problems. In its first year of operation, the plant developed vibrations in its fuel core and other malfunctions that forced the facility to shut down for more than a year. The repairs cost only $3 million, but Consumers had to spend an additional $6 million each month for replacement power purchased from other utilities.

No sooner did Palisades resume operation than a turbine failed, shutting it down again. Over the next two years, problems continued to plague the reactor. By the end of 1978 the plant had spent two-thirds of its more than six-year existence closed for repairs, and company officials announced that they might have to shut Palisades down for two full years for replacement of equipment.

Customers of Consumers Power didn't think much of their company's promises about "low-cost" nuclear energy. In the 10 years that followed groundbreaking at Palisades, they saw their rates increase six times. A kilowatt-hour that sold for 1.7 cents in 1966 cost 3.5 cents a decade later.

Consumers Power may be an extreme example of a utility stricken with nuclear troubles, but it illustrates the problems common to almost every nuclear plant built in the United States: dramatic cost overruns followed by poor performance. According to June Allen, a leading Virginia nuclear critic, "Even the most unlettered customer knows there's something wrong if great 'savings' from reactors double his electric bill."

There are indications that nuclear-related increases are just beginning. Commonwealth Edison's vice chairman Gordon Corey, whose company operates more nuclear plants than any other utility, has warned consumers, "We can't expect electric rates to be less than double in the 1980s what they are now, and probably more."

Officials of Consumers Power did not let problems at the Palisades plant dishearten them. As the plant lay idle, they were busy planning two more reactors. By 1977 cost estimates for their new Midland facility had risen from $349 million to $1.83 billion, even though construction had barely begun.

Why would a utility continue its commitment to atomic energy in light of its bad experiences with nuclear reactors? The answer lies in the peculiar environment in which the power companies operate.

As regulated monopolies, electric utilities are not subject

" YOUR CHECK, SIR ! "

Marlette. Copyright The Charlotte Observer.

to the normal pressures of competition. The "cost-plus" feature of the rate-making formula means that anything a utility buys, from postage stamps to power plants, can be charged directly to its customers without any sacrifice in profits.

In principle, utility rate making is a substitute for competition. Regulators attempt to set rates that allow a utility to recover its operating costs *plus* a profit on its investment, comparable to the profits earned by competitive businesses.

Three factors determine the amount of money that a utility is permitted to collect from its customers: operating expenses, rate base, and rate of return. Operating expenses are the day-to-day costs of doing business—items such as fuel, salaries, maintenance, and depreciation. The rate base is the amount of money a utility has invested in the production and marketing of electricity; it includes primarily tangible items such as power plants, transmission lines, and buildings. Public utility commissions allow the firms to recoup all their power-producing investments plus a profit, which is calculated as a percentage of that outlay. This is called a rate of return.

As the rate-making formula demonstrates, utility rates are set on a "cost-plus" basis: the commission calculates a company's operating costs during a given year and sets rates that allow the firm to recover a profit in addition to those costs. The size of a utility's profit depends on the size of its investment. Thus, the larger a utility's rate base, the larger its profit. The rate-making process obviously has a built-in bias toward expansion and capital-intensive technologies such as nuclear power. Many utilities have been criticized for "overbuilding" in order to "pad" their rate bases. Moreover, most private, investor-owned utilities—which provide 75 percent of U.S. electricity—have avoided serious energy conservation programs for the same reasons.

Because of the high cost of new power plants, rapid expansion is a great burden on utility ratepayers. As each new facility enters the rate base, a utility must increase its rates in order to maintain its profit levels. The process of utility regulation ensures that a utility's stockholders reap the benefits of investment in atomic energy, while its customers bear most of the risks.

Utility officials' decisions on when and how to build a new power plant are crucial to cost-conscious consumers. While construction will probably increase rates, the various power-generating methods affect utility rates in different ways. Both the cost of the plant and the cost of its fuel and maintenance must receive careful consideration. Nuclear plants are more expensive to build than conventional fossil fuel (e.g., coal or oil) plants of the same capacity. But they offer lower operating costs if they operate well, without accidents or costly shutdowns. Utility officials insist that when the costs are totaled, nuclear is less expensive. Nuclear critics disagree. The controversy over the economics of nuclear power rests on the proper assessment of these two very different kinds of costs.

Besides the rate-making formula, most regulatory commissions offer additional features that encourage utilities to "go nuclear." The most common of these is the fuel adjustment clause, which allows a power company to adjust its rates automatically according to its changing fuel expenditures.

Conceived as a measure to protect utilities from the uncertainties of rising fuel costs, the fuel adjustment clause has also served to insulate utilities from the risks associated with nuclear-plant operations. When a reactor is shut down or has its output cut unexpectedly, the utility must locate an alternative source of power. Most utilities can generate their own replacement power at an idle coal- or oil-fired plant but at a substantial increase in fuel costs. The fuel clause permits a utility to pass these increased costs to its customers immediately. Without this provision, the cost of replacement fuel would be absorbed by the utility's stockholders—at least until the next rate increase is permitted.

For a few utilities, the combination of nuclear power and automatic fuel adjustment has bestowed even greater benefits. Some fuel adjustments have been designed so that any increases in fossil fuel costs would be applied to all kilowatt-hours generated by the utility. Thus, the utilities could charge their customers for fossil fuel expenses on power from nuclear plants! According to the Pennsylvania Attorney General's office, Philadelphia Electric used this trick to earn an additional $1 million per month in 1975.

Utilities that cannot generate replacement power for their idle nuclear reactors must import electricity from other companies. Many have persuaded their regulators to establish purchased-power adjustment clauses, whereby rates may be adjusted automatically to reflect the cost of imported electricity. The Michigan Public Service Commission enacted this device in order to save Consumers Power from serious financial difficulties occasioned by the failure of its Palisades plant. Virginia consumer attorney

John Schell has suggested that fuel and purchased power adjustments should more appropriately be called "nuclear unreliability adjustments."

While some regulatory policies simply eliminate the risks associated with nuclear investments, others actually make those investments appear more profitable to the utilities. Several commissions, for example, have begun to allow utilities to include power facilities that are under construction in their rate bases, instead of permitting them to charge their customers only for completed facilities. This procedure, called construction work in progress, or CWIP, boosts electric rates by approximately 15 percent and increases utility profits as well. In fact, with CWIP, the more money a utility spends on construction, the more it can profit. CWIP also enables a utility to require its customers to provide capital for new construction, thus reducing its need to borrow money. Some Wall Street executives believe that future nuclear expansion will be difficult, if not impossible, without CWIP.

While most utility commissions have traditionally been devout proponents of atomic energy, in recent years a few regulators have begun to ask questions about the utilities' commitments to nuclear power. These commissioners are growing tired of granting repeated rate increases to cover reactor cost-overruns and operating problems. They are also concerned about future uranium prices and waste-disposal costs. Says former Michigan Commissioner William Ralls, "My experience seems to be that the planning and management problems of power production grow when nuclear projects are undertaken."

State utility commissions have immense authority and opportunities for dealing with nuclear power issues. In addition to deciding which utility expenses may be charged to customers, regulators commonly have jurisdiction over utility financing plans and when or where new power plants may be built. A growing number of commissions have begun to use these powers to "crack down" on nuclear investment.

In 1976, for example, the Pennsylvania Public Utility Commission prohibited two utilities from charging their customers for a $9 million construction error at a nuclear plant. And in that same year, the Maine commission, under pressure from citizens, ordered Central Maine Power to refund $3 million that it had charged its customers for replacement power when its reactor was closed for repairs. The Wisconsin Public Service Commission has gone a step further, ordering Madison Gas and Electric to cease expenditures on nuclear construction because of excessive costs. At this writing, the Pennsylvania Public Utility Commission has refused to allow Metropolitan Edison, the owner of Three Mile Island, to charge its customers for the costly accident at that plant. (This issue is treated in depth following this article.)

Utility commissions need not prohibit the construction of a reactor in order to discourage nuclear expansion. Several utilities have canceled nuclear plants because of "unfavorable" actions by their regulators. By eliminating adjustment clauses and other regulatory subsidies of nuclear power, commissions can force utilities to take responsibility for their decisions to build nuclear plants. The power companies will not be as willing to gamble with atomic energy when their own money is at stake.

Federal tax laws provide another regulatory subsidy for large investments such as nuclear reactors. Like other private businesses, utilities annually receive billions of dollars in tax breaks for investing in new equipment. An obscure clause in the federal tax code restricts state utility commissions from requiring power companies to pass their tax savings on to consumers. As a result, most of the nation's utilities collect millions of dollars in "phantom taxes" from their customers each year, which are never paid to the government. The more money a utility invests in construction each year, the more phantom taxes it can collect. Not surprisingly, the power companies with the largest nuclear construction programs are the leading phantom-tax collectors.

Financing construction is no easy task. Raising hundreds of millions of dollars can put a huge financial strain on a utility. But there are few financial problems that higher rates can't solve. Increased commission allowance of adjustment clauses, CWIP, and phantom taxes had made it easier for utilities to generate profits from their nuclear investments.

While rising capital and fuel prices are increasing the cost of nuclear power, the full costs associated with atomic energy have not yet appeared in electric bills. Uranium enrichment, for example, has been federally subsidized for more than 20 years; future costs, such as decommissioning and waste disposal, have simply been ignored. Nuclear power received additional subsidies through taxpayer-financed research and development and a federal exemption for utilities from liability in the event of a nuclear accident. Critics believe that consumers would reject nuclear power if they were required to pay the full costs of this energy source.

While it may be reasonable for the government to help private firms to overcome the uncertainties of a promising technology, after several years a new industry should be able to operate without a federal crutch. Yet after nearly 30 years, the federal Department of Energy still spends billions of dollars annually on nuclear research and development. One of the DOE's predecessors, the Energy Research and Development Administration, has conservatively estimated that the government has invested $9 billion in nuclear technologies. If this sum were allocated equally over the cumulative output of nuclear plants between 1957 and 1977, the cost per kilowatt-hour of nuclear research and development alone would be 1.1 cents. If the utilities had paid these expenses, the cost of nuclear-generated electricity would have increased by more than 50 percent.

Perhaps the most important subsidy of nuclear energy is the Price-Anderson Act, which protects utilities from liability in the event of a nuclear accident. This law provides utilities with federally guaranteed insurance for damages resulting from a nuclear accident and exempts them from liability beyond the amount covered by insurance. (This subsidy is discussed in detail in Senator Mike Gravel's article later in this volume.)

Armed with reams of press releases and multimillion-dollar ad campaigns, power companies go to great lengths to convince consumers that nuclear power saves them money. Fortunately, nuclear critics are beginning to pick apart the utilities' carefully worded statements about nuclear costs.

The nuclear industry's favorite tactic is to ignore everything but fuel costs. In 1976, for example, the Atomic Industrial Forum said in a statement to the press, "The contribution of nuclear power . . . represents a savings in generating costs of more than $2 billion" in 1975, which "translated into real savings for utility customers." The statement made no mention of capital costs, operating and maintenance expenses, or the costs of breakdowns. By this methodology, even the ill-fated Palisades plant appears to be four times cheaper than coal.

While nuclear proponents do not necessarily lie when limiting their analyses to fuel costs, they intentionally mislead the public. More often than not, newspapers fail to mention that fuel is only a fraction of the cost when they print utilities' claims of savings from nuclear reactors.

Utilities also commonly contrast nuclear costs to the most expensive alternative—oil—even though coal and energy conservation are usually more plausible alternatives. Moreover, when touting their nuclear savings, utilities often look only at the time periods when their reactors operate smoothly. Shutdown periods and expensive replacement fuel and purchased power costs are thus excluded from the record.

Ironically, the utilities have found a way to use their frequent nuclear plant malfunctions to make the "savings" from these facilities appear even greater. When a nuclear plant resumes operations after an outage, consumers often receive a sudden reduction in their rates as the utility ceases its use of expensive replacement fuel. Naturally, the utility attributes this cost savings to "low-cost nuclear energy." For example, when the Tennessee Valley Authority's Browns Ferry plant came back on line after a lengthy outage costing $240 million, the utility lauded its facility for helping to "hold down fuel cost and power purchases."

Extravagant claims of cost savings from a nuclear plant can be laid to rest by informed citizens. In 1975, for example, Consolidated Edison Company (Con Ed) blanketed New York City with news stories and bill stuffers claiming that its nuclear plants had saved consumers $95 million in fuel costs during the previous year. The Council on Economic Priorities (CEP) responded with a study showing that when capital and outage costs were considered, nuclear power's savings relative to oil were reduced to $4.3 million. CEP's analysis further indicated that rates would have been much lower had Con Ed built a coal-fired plant in the first place.

The American public was lured into the use of nuclear power by the promises of President Eisenhower and others that this new technology would provide a cheap and reliable form of energy. Relying on a multimillion dollar public relations campaign, the nuclear industry has managed to perpetuate the myth of cheap nuclear power for more than 20 years. But a string of nuclear-related rate hikes, coupled with costly accidents like the one at Three Mile Island, belie the power companies' claims. With the debunking of these myths, the nuclear industry's justification for existence has been shown to be nonexistent.

Who Will Pay for Three Mile Island?

In early April of 1979, when nervous investors were contemplating how to unload their stocks in nuclear-related businesses following the Three Mile Island nuclear plant accident, Wall Street analyst Mark D. Luftig of Salomon Brothers had a better idea.

Buy General Public Utilities, he told clients. At the panic price, he explained, bullish investors couldn't lose by buying stock in the company that owns Three Mile Island. As for the costs of the accident, he assumed they would be passed on to consumers. Indeed, Luftig claimed, such risks in a regulated industry had to be borne by ratepayers, or shareholders could not be induced to invest.

"It's costing the ratepayer . . . that's the way this business works," agreed George Troffer, a spokesman for Metropolitan Edison (Met Ed).

Both Luftig and Troffer were aware that, thanks to an automatic fuel adjustment clause, the rates of Met Ed and other General Public Utilities (GPU) electric companies would soon rise smoothly by as much as 25 percent to cover the cost of replacement power for the disabled Three Mile Island plant.

That cost of about $300,000 to $600,000 *per day* would simply be added on to the estimated $350,000 in daily charges that GPU customers are already paying for Unit

Two at Three Mile Island—whether or not it's working. The utility's Washington attorney, Gerald Charnoff, defended this arrangement on national television following the accident, saying, "It seems to me inescapable that considerations of fairness and ultimate economic impact require that the cost of replacement power be flowed through to customers."

But these are only the initial costs of the accident. Mopping up the radioactive water and cleaning the walls of the containment building alone is expected to cost $40 million. Then, depending on the degree of damage, the reactor will probably need a $100 million new core. Repair costs could prove so astronomical that the $780 million reactor would simply be scrapped. But even abandoning the plant would require decommissioning—removing or sealing off all radioactive components—which would cost an additional $50 to $100 million.

Although the utility's property insurance might cover up to $300 million of these costs, the rest would have to be absorbed by the ratepayers or by the company's stockholders. If Metropolitan Edison has its way, customers will pay whatever costs come along.

Pennsylvania regulators aren't so sure. Under pressure from its own staff and the office of the state consumer advocate, the Pennsylvania Public Utility Commission has dropped Three Mile Island's Unit Two from Met Ed's rate base, sparing customers from paying for the disabled plant, although they are still paying higher prices for replacement power.

Inability to charge customers for the cost of the accident could mean bankruptcy for Metropolitan Edison's parent company, General Public Utilities—one of the nation's largest power firms. Is bankruptcy unthinkable? In other industries where stockholders make mistaken investment decisions and allow management to run the business into

'Gentlemen, I don't think any of us really considered until now the ultimate danger posed by a nuclear accident . . . bankruptcy.'

Tony Auth. © 1979, The Philadelphia Inquirer. The Washington Post Writers Group.

the ground, the answer is no—or at least it used to be, before Congress saved the Lockheed Corporation from bankruptcy in 1971.

In the case of General Public Utilities, bankruptcy could conceivably occur without any interruption in service to customers. A court-appointed overseer would take charge of the company while finances were being shuffled; few if any jobs need be lost; and another utility could simply buy the GPU system on the auction block. Or GPU could be acquired by the State of Pennsylvania and turned into a state-owned system like New York State's PASNY. Or, several different localities might acquire pieces of GPU and run them as "appropriate scale" public power systems.

Securities analysts like Mark Luftig, of course, as well as the nation's utilities, would yell bloody murder if bankruptcy were allowed to occur. But the only alternative may be to reward Metropolitan Edison for greedily gambling with hundreds of thousands of human lives. Consumers had no control over what went on at Three Mile Island and are faced with a painful increase in their electric bills to pay for a piece of equipment that may never produce electricity again. One thing is clear: Someone will have to pay for it.

—Richard E. Morgan and Andy Feeney

If Nuclear Power
Is So Safe,
Why Can't It
Be Insured?

Senator Mike Gravel

The Price-Anderson Act, passed by Congress in 1957, insulates the nuclear industry from full financial responsibility for atomic accidents. The law sets a $560 million limit on industry liability in case of an accident—and part of that coverage would be provided by federal taxpayers. No one disputes that a serious nuclear mishap would kill tens of thousands of people, and cause billions of dollars of property damage.

Congress and the nuclear power industry face a basic and irresolvable paradox in trying to insure nuclear power, because the kinds of risks we take in generating nuclear electricity are simply not insurable.

There are several reasons. The most obvious is the scale of the possible accident. Here are the government's own various estimates of possible damages in a nuclear catastrophe: $6 billion, $7 billion, $17 billion, conceivably even

$280 billion; with outright deaths of 3400 to 43,000; and contamination of an area the size of California or Pennsylvania.

Other factors making nuclear power uninsurable are the far-reaching and long-term nature of the damage that could be done.

If great quantities of radioactive elements escaped into our environment, there would be no way to judge the extent of the cancer cases or birth deformities that might be caused. The health consequences would be incalculable —and recompense would thus be impossible.

Cancer, as in the case in Hiroshima, might appear 20 years *or more* after the event; Price-Anderson's statute of limitations is 20 years. And the range over which damage would be inflicted cannot be judged. Radioactive carcinogens would be distributed throughout the biosphere and concentrated by some plants and animals in a way completely outside the control of man.

Those familiar with the issue already know that nuclear power is uninsurable. If it were insurable, nuclear power would be subject to normal liability laws, and there would be no need for Price-Anderson. Perhaps the real decision is, put very bluntly: "Who's going to be left holding the bag?"

If the Indian Point reactor, 24 miles from New York City, suffers a meltdown and a breach of containment, who's going to be left holding the bag? If the San Joaquin Project contaminates this nation's largest food-producing area, who's going to be left holding the bag?

Back in the mid-1950s, when the insurance problem was first tackled by Congress, the nuclear industry made it very clear who was *not* going to be left holding the bag. They let us know that if normal liability were allowed to apply, they would not take the nuclear gamble. The public, on the other hand, was not wary. And today, it is the public that really carries the burden of nuclear power risks, even though few citizens are aware of it.

Congress decided that in the worst nuclear catastrophes, the citizen's constitutional right to just compensation would be suspended. An arbitrary sum—$560 million, a fraction of the possible damages—would be shared on a pro-rated basis. And even if victims get only pennies to the dollar under this scheme, Price-Anderson prohibits them from receiving additional compensation from the nuclear industry.

To add insult to injury, some of the $560 million insurance is provided by the federal government. The government charges less for its coverage than private insurers charge for theirs. Federal insurance was arranged because private insurers refused to supply even the $560 million of required insurance. This refusal is important; it is strong evidence that the insurance industry doesn't believe claims of nuclear safety. If it did, it would be anxious to sell all the insurance it could. Insurers like to collect premiums for coverage of claims resulting from accidents that will never happen. The lack of confidence in nuclear safety by the insurance industry—our best risk assessors—should serve as a strong warning of nuclear dangers.

This lack of confidence is further demonstrated by the fact that insurance companies explicitly exclude coverage of nuclear disasters from their homeowners and other insurance policies.

So, in a word, the public is holding the bag.

Or, as a Columbia University study of Price-Anderson declared in 1974: "The decision to limit liability represents a determination that a major share of the costs of an accident should be borne by its victims."

I believe we have failed our obligation to the public in permitting the limitation of corporate liability. The law says that should damages exceed $560 million, Congress may provide disaster relief. But another Columbia study has shown that congressional action in such instances can

Credit: Environmental Action Reprint Service.

be expected to be too little, too late. And if congressional action is what's needed to cover a nuclear catastrophe, then what purpose does Price-Anderson serve? Nothing in the act guarantees extra relief.

The only conclusion is that Price-Anderson exists to protect the utilities and nuclear manufacturers. Period.

Even with those provisions that supposedly assure $560 million for victims, there are loopholes that protect industry at the expense of the public. For one thing, industry's costs for investigating and settling claims are to be subtracted from the $560 million. In other words, the funds to pay utility lawyers to challenge victim's claims will come from money supposedly designated for paying those claims. In addition, the utility's property, outside the reactor, is covered by the $560 million. That is, the utility will recover from its own "public liability" insurance.

Admirers of the Price-Anderson Act point to its assurances that money will be available to victims on a no-fault basis. In fact, however, the guarantees made by Price-Anderson are illusory. To begin with, in many states utilities would be held strictly liable even without Price-Anderson. In addition, victims must accept the settlement offered to them or else go to court much as though Price-Anderson's no-fault provisions did not exist. How many people are likely to accept their pro-rated share if the damages exceed $560 million?

It should also be noted that the $560 million provided by Price-Anderson is now worth much less than it was when it was settled on over 20 years ago. And in that same period, the possible consequences of an accident have grown, because reactors are larger and the population density near them greater.

Finally, there is the problem of proving damages. Even under the no-fault provisions, victims must prove that their injuries were caused by the accident. But a cancer does not grow with a flag in it identifying its cause. It is impossible to prove that any individual cancer has been caused by radiation. And the industry is not going to give victims the benefit of the doubt.

The case of Edward Gleason, a truck dock worker, is instructive.

In 1963 Gleason handled an unmarked, leaking case of plutonium that was being shipped to a nuclear facility. Four years later he developed a rare cancer: his hand and then his arm and shoulder were amputated. He sued, but even though plutonium is among the deadliest of radioactive poisons, company insurers said his plutonium accident could not be proved as the cause of the cancer. Gleason's suit was dismissed in 1970 on statute-of-limitations grounds, although a settlement was later made. He died of cancer in 1973 at the age of 39.

Finally, even if the victim can prove his case, how much of the $560 million is going to be left for those who suffer cancer years after the accident? Only by restoring the citizen's right to recover relief beyond the no-fault amount can Price-Anderson claim to treat the public fairly. And the nuclear power industry should be exposed to such suits, not isolated from the consequences of nuclear accidents.

We should also be aware that Price-Anderson has acted as an industry subsidy. As such, it has made nuclear power look more attractive economically than it really is. This has resulted in pouring tens of billions of dollars into nuclear fission and keeping those dollars from the benign and renewable energy sources that we should have taken seriously long ago.

The most important point remains to be made.

The reality of a nuclear catastrophe exceeds the capacity of the imagination: a hundred thousand people each suffering $60,000 in damages in the case of the $6 billion estimate; or a million people each suffering $17,000 in damages in the $17 billion case. There has never been a peacetime disaster on such a scale.

As long as we rely on nuclear power, we court this kind of disaster. There is only one way to protect the public from the consequences of a nuclear power catastrophe— stop building nuclear power reactors.

There is no more graphic example of the hypocrisy of nuclear promoters' arguments than Price-Anderson. On the one hand they tell us major nuclear accidents won't happen. Then they lobby vociferously to protect their assets from such accidents! One hardly needs to be a nuclear physicist to understand this issue.

Harold Green, an attorney and law professor, has stated the Price-Anderson paradox most precisely. He wrote in the Michigan Law Review:

> The fact that the technology exists and grows only because of Price-Anderson has been artfully concealed from public view so that consideration of the indemnity legislation would not trigger public debate as to whether nuclear power was needed and whether its risks were acceptable. . . .

It is remarkable that the atomic energy establishment

has been so successful in procuring public acceptance of nuclear power in view of the extraordinary risks of the technology that are so thoroughly and incontrovertably documented by the mere existence of [the Price-Anderson Act].

If utilities choose to gamble with nuclear power, let them share the risks along with the rest of us. If the odds are really as favorable as the industry advertises, the risk should not bother the utilities.

If utilities will not build nuclear power plants under these conditions, then it is time for us to confront the *meaning* of their refusal.

In the interests of launching the nuclear power industry, the constitutional right of the citizen to just compensation was suspended. The normal deterrent to reckless activity —financial liability for the consequences—was suspended. This abridgment of citizen rights and business prudence was not justified in the early days of the industry, and it is even more offensive today.

In March 1977 Judge James B. MacMillan of North Carolina declared Price-Anderson's limit on liability unconstitutional. A year later the U.S. Supreme Court overturned his decision, noting that Congress had repeatedly demonstrated its support for nuclear power including passage of Price-Anderson itself. This puts full responsibility back on Congress to repeal the limit on liability and subject nuclear power to a more realistic risk-versus-benefit calculation by investors.

The Sun Offers Energy and Jobs

Douglas Fraser

Today, when our nation needs both safe energy and good jobs, solar power can provide both.

Enough solar energy reaches the United States in 12 daylight hours to equal the nation's yearly energy consumption, according to the Department of Energy. A fully developed solar industry tapping this safe, embargo-proof energy source, accompanied by a national commitment to high energy efficiency, would be able to meet much of the country's energy requirement. Properly planned for and developed, such an industry would stimulate economic growth while helping to preserve the environment and safeguard public health.

In only a few years, our embryonic solar energy industry could become a major force in our economy, fueling our factories and heating our homes, schools, and businesses. A thriving, diverse, and extensive solar industry would greatly reduce the need for oil imports, provide jobs for millions of Americans, and move this country closer to its goals of energy independence and full employment.

A long stride toward the solar society of the future was taken with the celebration of Sun Day on May 3, 1978.

Many of its organizers were the devoted disciples of Earth Day, which had launched the environmental protection movement eight years earlier. Sun Day activities were supported by a diverse coalition of union leaders and workers, farmers, politicians, academicians, students, consumer and environmental activists, minority group representatives, and the urban poor. Demonstrations emphasized the viability of solar energy and sought to make its full development an immediate national priority. The United Automobile Workers (UAW) union was one of the sponsors of Sun Day, just as it was a sponsor of Earth Day in April 1970.

An attractive feature of the sun, from the standpoint of the consumer, the entrepreneur, and the solar technician, is that its energy is already evenly distributed. Technologies to capture sun power and make it work for us are being employed minimally now in America and around the world. Moreover, the technologies lend themselves to increased development and installation with local capital and local skills.

What are these technologies?

● *Passive solar construction:* designing and placing buildings, windows, and insulation to get the most heat and light from the sun.

● *Water and space heating:* using solar plate collectors on roofs and outside walls to provide heat and hot water.

● *On-site electrical conversion:* using photovoltaic cells to change sunlight directly into electricity, without turbines or electrical grids.

● *Wind energy systems:* using machines of various designs and sizes to harness the wind (created indirectly by the sun through temperature gradients), thereby generating electricity, pumping fluids, and compressing gas to do work.

● *Biomass conversion systems:* converting various types of organic matter (i.e., products of the natural food chain, which is fueled ultimately by photosynthesis and the sun) into liquid or gaseous fuels.

As long ago as 1952, the Paley Commission reported to President Truman that solar energy could play a more beneficial role in meeting energy needs than nuclear power. In fact, the Paley Commission believed that an aggressive effort at that time would have led to the use of solar power in heating 13 million homes and commercial buildings in the United States by 1975. Such an effort, of course, was never made, but not because we didn't know how to make it. Energy seemed cheap and plentiful in those days and few people believed there was any need to hurry solar power along.

But since the Arab oil embargo of 1973 awakened Americans to the nation's energy deficiencies, there has been rapid growth in solar energy *thinking*, and more important, in solar energy *doing*. As a result, the applied solar energy business has been doubling almost every nine months since then. Current projections of a Mitre Corporation study made for an AFL-CIO union, the Sheet Metal Workers International Association (SMWIA), point to at least a $10 billion solar industry by 1985. Dr. Henry Marvin, director of solar energy for the Department of Energy, predicts that by the year 2001 the solar heating and cooling industry in the United States is likely to be as large as the conventional heating and cooling industry is today. Commenting on the solar growth rate, the *New York Times* has reported that the solar industry is "progressing faster than either Washington or its advocates expected. . . . It seems likely to become *the* growth industry over the next two decades."

Solar heating technologies were competitive with electric resistance heating in 12 of 13 cities analyzed by the Federal Energy Administration, a forerunner of the Department of Energy. And great progress has been made in the production of photovoltaic cells—individualized energy systems with no moving parts and no radioactive wastes—making this form of energy conversion the only one whose price has been declining in the past few years. The President's Council on Environmental Quality is optimistic about this trend, having found that "there appear to be no fundamental barriers to further lowering of costs." If this proves true, the massive commercial use of photovoltaics will surely come about.

The nation would benefit enormously from these solar technologies; the fuel is free and renewable and they pose virtually none of the dangers associated with nuclear power. In addition, solar systems are smaller, less complex, and less expensive than the massive power systems favored by multinational energy companies. Communities can

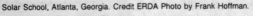

Solar School, Atlanta, Georgia. Credit ERDA Photo by Frank Hoffman.

build solar systems as needed and can add units as they see fit. By doing so, communities need not tie up billions of dollars over long planning and construction periods. Just as important, building and maintaining large numbers of diverse solar energy systems around the country would mean a steady supply of useful, important jobs in trades and professions employing both skilled and unskilled workers.

An estimated 25 percent of the funds spent on new solar heating and cooling systems would go to pay the labor costs of making and installing collectors, storage tanks, ducting, and assorted hardware. Edward Carlough, president of the SMWIA, says: "Even figured conservatively, energy-saving modifications with an expanded use of solar energy could put all unemployed sheet metal workers back to work."

A 1975 study by the SMWIA found that modest activity in heating and cooling alone would bring in more than one-quarter of a billion dollars in wages to those metal workers by 1990.

Moreover, the jobs would not be only for sheet metal workers. Employment would be provided for carpenters; cement masons; electricians; painters; plumbers; air-conditioning, heating, and refrigeration technicians; glaziers; crane operators; teamsters; and laborers. A program to retrofit all American households with solar water and space heating over the next 20 years would generate over one-half million jobs per year in installation alone, the President's environmental council estimates. Demand for copper, steel, aluminum, glass, and insulation would increase employment in those industries as well. Thus, jobs would also be generated for such skilled workers as machinists, precision-tool operators, surveyors, ironworkers, welders, and electrical and mechanial engineers. The Congressional Office of Technological Assessment says, "if all conventional power were replaced with solar units, labor requirements would be multiplied by a factor of two to five."

Additionally, analysis by the Bechtel Corporation and the University of Illinois shows that the number of indirect jobs created by new solar energy development would be about equal to the number of direct jobs.

Studies comparing the impact of solar and nuclear energy development on employment show that solar energy utilization provides more jobs:

• Fred Dubin, president of the engineering, planning, and management firm of Dubin, Mindel, Bloome, Associates, New York City, which has conducted comprehensive energy analyses for many regions of the country, found that $2 billion invested in energy conservation plus solar systems will provide four times as many jobs as when the same $2 billion is invested in nuclear power plants.

• The California Employment Development Department, in a preliminary study on solar energy in California, found that "the development of solar energy will require much more employment than other energy sources, in particular, electric power plants." According to this analysis, almost seven times more person-years of work (direct and indirect, for construction and operation over a 20-year period) would be generated if solar hot water and space heating were utilized instead of an equivalent amount of energy from nuclear power plants. Average wages paid through the solar alternative would be equal to or about the same as the nuclear route. The study also points out that the solar jobs would be located in populated areas, not in remote locations where increased energy activity would put "significant strains" on local services, especially on housing.

• Wind energy, according to Lee Johnson of *Rain* magazine, Portland, Oregon, needs only 42 percent of the money needed by atomic facilities to employ the same number of workers.

● A report ordered by the New York State Legislative Commission on Energy Systems calculated that the operation and maintenance of a large wind system requires two to four times more labor on a continuous basis than does a nuclear fission system.

Employment prospects do not end with the potential American market. The Agency for International Development (AID) and the World Bank are exploring the use of solar technologies in developing countries. It does not take much imagination to look at our energy-hungry world and speculate just how much of an international solar energy market there would be once we had the products efficiently produced by American workers: We would see additional jobs created in sales, advertising, wholesale and retail trade, shipping, government, law, and accounting—all involved in building up and maintaining a vital industry that would be around as long as the sun.

While it is clear that solar energy development can stimulate our economy, there will be difficulties in accomplishing a national transition to a solar society. Many people remain unconvinced of solar energy's potential in the near future and are reluctant to give up conventional ways of thinking about providing energy. For example, many large corporations may want to realize a return on their investments in fossil fuels or nuclear power before they are willing to venture into harnessing power from the sun. And there will be many administrative problems connected with such a massive conversion.

We must also face another problem: Although solar energy will generate millions of jobs as we convert from conventional fuel technologies, some American workers could lose the jobs they now hold unless considerable care is exercised. Under our economic system, it has always been the worker and his and her union that have borne the economic burden of technological change. Every effort must be made to ensure that this does not take place. Plans must be made to train and retrain sufficient people in solar skills and to make sure that those who lose jobs in conventional energy industries are not left out in the cold. Organized labor will accept nothing less than a well-planned switchover to solar power, which includes protection of the income, equity, and benefits of men and women working today. Although some environmentalists have not paid sufficient attention to this part of the energy-jobs equation, it is gratifying to see more and more of them addressing basic concerns about jobs and job security and preparing to take supportive action. The way Sun Day's organizers have reached out to organized labor is an excellent example of this.

The solar energy technologies discussed here are so diverse—and embrace such a wide range of skills and abilities—that opportunities will abound for all.

There is much that Americans can do to bring about the appropriate kind of transition. Currently, as the President's Council on Environmental Quality noted, substantial subsidies and advantages exist for nonsolar energy sources: tax credits for fossil energy producers and reduced insurance assessments for nuclear plant owners, for two examples. To balance the slate, barriers to commercialization of known and proven solar technologies must be removed. Many states already provide tax incentives to homeowners for solar installations, and the positive impact on the industry would be enormous if the federal government guaranteed solar energy markets.

The concept of Sun Day was a nationwide plan "to lead the United States into the solar era" and to show that putting sun power to work is not science fiction, not a plaything of the affluent, and not a curious art form. Attesting to the feasibility of its use, the UAW has installed wind and solar equipment to heat the Walter Reuther Family Education Center building in the Black Lake area of northern Michigan.

The growing solar movement reflects a basic confidence expressed by the Oregon Energy Council a few years ago when it said, "A transition to a solar economy is desirable and realizable. It involves neither privation nor social deprivation. Lifestyle changes would be minimal. The rewards would be enormous."

Yes, the rewards would be enormous. But first we have to make the effort. From almost every standpoint, solar energy comes up smiling—just as the sun has done for millions of years and will continue to do.

Getting More from Less

Denis Hayes

More than one-half the current U.S. energy budget is waste. For the next quarter century the United States could meet all its new energy needs simply by improving the efficiency of existing uses. The energy saved could be used for other purposes and relieve us of the immediate pressure to commit enormous resources to dangerous energy sources before we have fully explored all alternatives. Energy derived from conservation would be safe, more reliable, and less polluting than energy from any other source. Energy conservation could reduce our vulnerability in foreign affairs and improve our balance-of-payments position. Moreover, a strong energy conservation program would save consumers billions of dollars each year.

In 1975 Americans *wasted* more fuel than was *used* by two-thirds of the world's population. We annually consume more than twice as much fuel as we need to maintain our standard of living. We could lead lives as rich, healthy, and fulfilling—with as much comfort, and with more employment—using less than half of the energy now used.

These conclusions rest upon conservative assumptions. They assume that life styles will change only cosmetically

—that Americans will continue to travel as many miles, keep their homes just as warm, operate as many appliances, and eat what they now eat. If we consider functions instead of technologies—ways, for example, to eliminate the need for a certain type of transportation instead of merely ways to make it more efficient—the results will be more dramatic.

In the context of great uncertainty over future energy supplies, fresh attention is being paid to the potential for energy conservation. Most discussions proceed from economic arguments. And the economic arguments for conservation are strong. Energy prices have been pacing our spiraling inflation, and saving energy saves money.

While the financial community claims that the country verges on a major capital crisis, most estimates of the capital requirements for projected new energy sources range between $500 billion and $1 trillion by 1985.

A barrel of oil saved is more valuable than a new barrel of oil produced, due to the "energy cost" of production. A dollar invested in wise energy conservation makes more net energy available than a dollar invested in developing new energy resources. For example, ceiling insulation in a typical home costs about $300 installed and will save about seven barrels of oil each year for the lifetime of the house. Using very conservative discount rates, the present value of the energy to be saved by the insulation in future years is 60 barrels. Thus we are "producing" heating oil at about $5 per barrel when we install ceiling insulation. If heating oil costs only $3 per barrel, the insulation will not be economically attractive. But today heating oil costs $16 per barrel. Regulated natural gas costs $11 per barrel-equivalent, and electricity costs as much as $35 per barrel-equivalent. Ceiling insulation is, by comparison, dirt cheap.

At the individual home-owner level, home insulation may well guarantee a higher rate of return than *any* other investment available to the average citizen. Common stocks, corporate bonds, and savings accounts pay interest rates of 5 to 10 percent—often at some degree of risk. Home insulation may earn 20 to 40 percent in saved fuel costs at no risk. Moreover, investments in home insulation will raise property values.

Other examples of the economic advantages of conservation abound. The total cost of electricity from a large nuclear power plant is currently around $3000 per delivered kilowatt. This includes the cost of the power plant and the distribution system, adjusted for capacity factors and line losses. It also includes conservation figures for federal research, development, and subsidization of the fuel cycle; interest charges; fuel; radioactive waste management; regulation, etc. By comparison, replacing electrical resistance heating with an efficient heat pump costs between $50 and $120 per thermal kilowatt. Recapturing waste heat from the chimneys of industrial furnaces costs about $70 per thermal kilowatt. Generating electricity as a by-product of industrial steam production costs $190 to $280 per thermal kilowatt. The economic advantage of such conservation, especially in a period of general capital shortages, speaks for itself.

Energy conservation by Americans today will allow the earth's limited resource base of high-quality fuels to be stretched further. It will enable our children and those in other lands to share in the Earth's finite stock of fossil fuels. It will make an especially critical difference to those living in underdeveloped lands, where the marginal return per unit of fuel is far greater than in highly developed countries.

Energy conservation will allow a portion of the fossil fuel base to be reserved for nonenergy purposes: drugs, lubricants, and other materials. The energy cost of manufacturing such substances from carbon and hydrogen, once our present feedstocks have been exhausted, will be astronomical.

Power Lines. Photo by Daniel S. Brody.

Energy conservation will allow us to minimize the environmental degradation associated with all current energy-production technologies. The consequences of pollutants such as heavy metal particulates, carcinogenic aromatic hydrocarbons, and various radioactive materials could be terrifying. Energy conservation will allow us to avoid objectionable energy sources while the search for safe, sustainable sources continues. Conservation also decreases the likelihood that we will cross climatological thresholds (e.g., with carbon dioxide production or with regional heat generation), triggering consequences that may be devastating.

Energy conservation will contribute to human health. Much of the fat in our energy diet leads to fat on our bodies. Energy conservation could lead to more exercise, better diets, less pollution, and other indirect benefits to human health.

The security of a modest energy supply is more easily assured than that of an enormous one that depends upon a far-flung network of sources. And an enlightened program of energy conservation will substantially bolster employment levels.

Energy conservation is an unequivocally desirable national objective and already among the most widely supported goals in the country. The time has come to translate all those supportive words into supportive action.

However, there are barriers to such action. The United States matured in an era of abundant fuel and declining real-energy prices. Energy was substituted for all other factors of production—including, wherever possible, human activity. We became a sedentary society, trading vast amounts of cheap fuel to avoid trifling exertions or momentary inconveniences. For example, to avoid occasionally moving their feet 10 inches and their hands 6 inches, most American car buyers pay extra for automatic transmissions that decrease gasoline mileage by 10 percent or more. Often (as with our national habit of leaving on unnecessary lights) we squander energy on nothing.

Our energy growth rates are neither "accidental" nor "natural"; they have resulted from conscious decisions. Energy consumption levels have been pushed, pulled, shoved, and kicked upward by every trick and tactic known to the contemporary science of mass marketing. To encourage growth, fuel prices have been kept artificially low and utility rate structures have rewarded waste. Environmental costs and health costs have been ignored.

Our relentlessly rising fuel consumption has had an institutional rationale. A dollar invested in facilities to produce more energy makes energy available to the producer, who then sells it for profit. Although the same dollar invested in conserving energy often makes far more energy available, conserved energy (which would otherwise be wasted) is energy that the energy producer had already counted as sold; the company, for whom a dollar burned is a dollar earned, is generally unenthusiastic about "returned merchandise." If a utility sells a billion kilowatt hours this year, and 10 years from now is still selling a billion kilowatt hours, its dividend-conscious stockholders will take little satisfaction in the greater efficiency and benefits of the future billion. Corporate officers cannot relish the prospect of informing stockholers and lending institutions that their company has completed a successful transition into a no-growth economy.

The U.S. energy industry, with its unrivaled resources, is a growth industry par excellence. No countervailing institution with comparable muscle exists to promote energy conservation. Some corporations sell energy conservation programs, but these are invariably marginal operations by firms whose dominant interests lie elsewhere. Governmental agencies with a conservation mandate chronically lack funds and authority.

Our institutions of commerce and government, fashioned to serve the cause of growth, have succeeded admirably. But while they know full well how to grow, they don't know how "not to grow." Like animals bred strictly for size, they find themselves confronted with unexpected vulnerabilities.

A true national debate about our real-energy options has never taken place. The public continues to be told that the relationship between energy use and well-being is one-to-one. Adherents to this viewpoint proclaim: "The more energy we use, the better off we are. If we want to be even better off in the future, we will have to use even more energy. Energy conservation would mean a poorer America."

In its "Energy to the Year 1985" report, the Chase Manhattan Bank expresses this viewpoint forcefully:

> It has been recommended in some quarters that the United States should curb its use of energy as a means of alleviating the shortage of supply. However, an analysis of the uses of energy reveals little scope for major reductions without harm to the nation's economy and its standard of living. The great bulk of the energy is utilized for essential purposes—as much as two-thirds is for business related reasons. And most of the remaining third serves essential private needs. Conceivably, the use of energy for such recreational purposes as vacation travel and the viewing of television might be reduced —but not without widespread economic and political repercussions. There are some minor uses of energy that could be regarded as strictly non-essential—but their elimination would not permit any significant savings.

This statement and others like it made by the energy industry and its financial backers confuse energy conservation and curtailment. Curtailment means giving up automobiles; conservation means trading in a 7-mile-per-gallon status symbol for a 40-mile-per-gallon commuter vehicle. Curtailment means a cold house; conservation means a well-insulated house with an efficient heating system. Energy conservation does not require the curtailment of vital services; it merely requires the curtailment of energy waste.

Yet "waste" can mean different things to different people. Waste signifies one thing to a physicist and another to an economist; it has widely differing meanings for philosophers, engineers, and politicians. In fact, all energy policy discussions bear this curse of Babel; they are plagued by ambiguous terminology and consequent misunderstandings. Energy policy, a new and eclectic field, involves so many diverse disciplines that a common language and set of definitions could hardly be expected. But many conflicting claims might well be reconciled if only their respective proponents were talking about the same thing.

Few energy economists, for example, have any background in thermodynamics. Few know that energy has a qualitative dimension, that the Second Law of Thermodynamics—which states that the quality of energy declines as it is used—is just as absolute as the First Law, which states that the quantity of energy in the universe is constant. Most studies of energy use have dealt only with the quantitative dimension of energy. Most have considered the flow of energy units (Btu's, calories, or joules) used in a given process but have not distinguished between the relative entropy levels (i.e., levels of organization and quality) of these quantities. Most have thus ignored the most important aspect of the energy flows they have been analyzing.

One distinguished team of scientists has expanded the boundaries of the energy discussion to cover quality as well as quantity. Using the concept of "free energy" as developed in 1878 by J. W. Gibbs as the basis of analysis, physicists under the auspices of the American Physical Society have scrutinized many of our uses of energy. Introducing the valuable concept of "second law efficiency," the study concludes that the overall "second law" efficiency of energy use in the United States is between 10 percent and 15 percent. Cars are found to be 10 percent efficient, home heating 6 percent, air conditioning 5 percent efficient, and water heating only 3 percent efficient using the second law criterion.

While physicists thus argue that energy use in the United States is only 10 to 15 percent efficient, many economists believe that there is no significant waste in our present energy budget. By their own standards, both camps are correct. The physicists failed to examine the economic cost of increasing the physical efficiency of energy use. Nor did they examine systemic alternatives (e.g., substituting van pools or public transit for automobiles). Theirs was a purely technical study of the efficiency of use of free energy in current technologies. Most economists, on the other hand, disregard the physical and technical phenomena their idealized marketplaces purportedly represent. They take for granted that pricing mechanisms have assigned appropriate dollar values to all possible purchases. Since fuel buyers act in their own economic self-interest, and since the total economy seems to be operating reasonably efficiently, these economists argue that our current level of fuel consumption cannot be considered economically wasteful.

If both perspectives are "correct," both have shortcomings. In economic terms, technical opportunities for conservation mean little if they are prohibitively expensive. On the other hand, the purely economic perspective may be even more deficient. Its guiding principle—that a dollar should be invested wherever it will bring the highest return—is sensible for many purposes. However, at present it almost completely disregards such "externalities" as environmental quality, occupational safety, and national security. Moreover, it ignores the needs of the next guy in line. On a planet with rapidly depleting, finite resources, future generations can't fend for themselves; the economic principle must be tempered by humanitarian constraints. But economics is an analytical tool, not a system of ethics.

Combining the insights of both physicists and economists, this author considers energy to be "wasted" whenever work is performed that could have been completed with less or lower-quality energy and without incurring higher total social or economic costs. By this definition, the United States consumed about twice as much fuel in 1975 as was necessary. The major areas in which significant savings could be made are transportation, heating and cooling systems for buildings, water heating, the food system, electrical generation, industrial efficiency, waste recovery, recycling, lighting, and transportation.

Transportation presently accounts for about 24 percent of our direct fuel consumption. Another 18 percent of our energy budget is used indirectly—to build and maintain vehicles, construct roads, etc. Sixteen percent of our direct fuel consumption and an additional 6 percent of our indirect consumption could be saved by gradually tripling the mileage performance of individual vehicles, substantially reducing average vehicle size, transferring half of commuter traffic to multiple-passenger modes while reducing the number of automobiles accordingly, and systematically shifting freight to more efficient modes. These savings could be phased in over the next 25 years.

Moreover, with transportation accounting, directly and indirectly, for 42 percent of our total energy budget, we need to ask why and where we are going. If we diminish our volume of travel, even greater savings will result.

Space heating currently comprises 18 percent of our energy use, water heating 4 percent, and air conditioning 3 percent. The combination thus totals 25 percent of the nation's direct fuel consumption. Strict insulation stan-

Solar rooftop collectors being installed on an energy-saving home. Credit: Department of Energy.

dards on new buildings, a vigorous program to increase insulation in existing residential and commercial buildings, the use of solar heating and cooling technologies (cross-bred where appropriate with compatible technologies like heat pumps), the adoption of solar water heaters for virtu-ally all residential and commercial hot water needs, and the widespread use of total energy systems in large complexes could reduce our energy budget by at least 16 percent over the next quarter century.

The U.S. food system uses more than 12 percent of the nation's fuel budget; farming uses 2½ percent; and food processing, packaging, retailing, and preparation account for the remainder. Improvements can be made at every stage of the food system—especially in processing and preparation. Technical improvements could save 3 percent of the national energy budget; home gardens, improved diets, and a switch away from fast foods could save an additional 2 percent.

About 25 percent of the U.S. fuel budget goes into electricity generation, of which approximately one-third actually emerges as usable power. Two-thirds of the potential energy fed to power plants is cast off as waste heat. Alternative generating technologies currently being developed will be able to operate at much higher efficiencies. For example, laboratory-size fuel cells have operated at 75 percent efficiency, and combined cycle power plants can obtain a 55 percent efficiency. By increasing electrical conversion efficiency to 45 percent, we can save 3 percent of our fuel budget.

Our current electrical price rates were designed in an era of 8 percent annual electrical growth. Since they do not generally fluctuate with the time of the day or with the changing seasons, enormous demands are made on electrical generating facilities at three o'clock on a July afternoon, while the same generating system may be mostly idle at three o'clock on a February morning. Rates that were scaled according to gross demand would trim the differences between the peaks and valleys, thus reducing the amount of capital tied up in mostly unused generating capacity. Furthermore, electricity prices have been designed so that the largest users pay the lowest rates. While this once seemed reasonable (in terms of cost of service), such rate schedules have outlived their rationale and now serve largely to perpetuate inefficiency. Flattened rates, or inverted rates (in which heavy users would pay more per unit) would substantially cut the total electrical demand. Already public utility commissions in several states are experimenting with electrical rate design.

Sweden manages to harness approximately one-third of the waste heat from its power plants for commercial purposes; the United States recaptures essentially none. Utilizing just 12 percent of this waste heat, we could reduce our fuel demand by 2 percent. Total energy systems are being designed toward this end for use in U.S. hospitals, shopping centers, and other large complexes. Similarly, the high-quality energy now used to produce industrial (low-grade) steam could be used to generate electricity during the conversion process. Generating 50,000 megawatts of electricity in this manner is economically attractive right now. As industrial electric rates increase rapidly in the years ahead, the economic attraction of this power source will grow commensurately.

Other possibilities for industrial conservation abound. Many countries use more energy-efficient processes than the United States does, for industries ranging from the refining of ores to the manufacturing of final products. Important savings have been made in the last few years in the U.S. chemical industry (which uses one-fifth of all industrial energy) and the primary-metals industries (which use another fifth).

Much usable energy is currently thrown away. If we were to extract the optimum level of potential energy from urban refuse, human excreta, agricultural residue, feed-lot wastes, and forest-product wastes, and if we were to recycle 13 million tons of ferrous metals and 500,000 tons of aluminum, substitute standardized returnable bottles for most cans, and eliminate all the unnecessary packaging that so clutters up our lives, we could obtain more than 4 percent of our current energy needs.

One-half of the 4 percent of our direct energy now spent on lighting is superfluous, and most "necessary" lights operate inefficiently. Currently, incandescent bulbs convert only one-twentieth of the energy in electricity into light; fluorescent bulbs convert over one-fifth. As a NATO-sponsored scientific committee on energy conservation reported, however, there is "no fundamental theoretical reason why a 100 percent conversion efficiency" of electricity into light should not be attained.

No short list can exhaust the possibilities for substantial conservation that permeate our entire way of life. Most of the goods we purchase could be both manufactured and made to work more efficiently. All levels of government, for example, are larded with extraordinary waste. The largest single user of energy in the United States is the federal government, and the largest user inside the government is the Department of Defense. The Pentagon consumes 2 percent of the nation's fuel directly and another 4 percent of the entire national energy budget indirectly.

International comparisons support the contention that the 1975 U.S. energy budget can be trimmed over time by more than one-half. For example, in 1975 Sweden, West Germany, and Switzerland, with about the same level of per capita GNP as the United States, used only 60 percent as much energy per capita as the United States. West Germany used seven-eighths as much fuel per capita as the U.S. for industrial production, one-half as much for space heating, and only one-quarter as much for transportation. None of these countries has begun to approach its full potential for energy thrift.

The contention that the U.S. energy budget can be gradually cut by more than one-half without altering the nation's standard of living is almost certainly conservative. Nevertheless, such a reduction is neither inevitable nor even very likely. Even when clear goals are widely shared, they are not easily pursued. Policies tend to provoke opposition. Unanticipated side effects almost always occur. And sometimes we just don't know how to achieve certain ends. The praiseworthy goals of the Great Society have gone unrealized. The current American recession is not the handiwork of secret fans of zero economic growth.

Our past is one of persistent energy growth, and the past is widely presumed to be prologue to the future. This presumption guides the elaborate computations of most modern forecasting, and it underpins much of conventional wisdom. But as René Dubos has written, "Trend is not destiny." Calamities and bonanzas can intrude upon the smooth curves of extrapolation; people and nations can rethink their direction and alter course. Most standard energy forecasts are projections. They are judgments about tomorrow made today from data generated yesterday. They necessarily reflect a particular set of values and assumptions. Most have been unremarkably similar. All are likely to prove grievously flawed.

For students of energy policy, the future is not what it used to be. Consumption patterns for commercial fuels, after two decades of unbroken exponential growth, have changed radically over the last two years. Even more fundamental discontinuities seem likely in the near future. Momentous conflicts loom between habits and prices, convenience and vulnerability, the broad public good and narrow private interests.

The most striking opportunity now presenting itself lies in energy conservation. In the past, conservation was viewed as a marginal activity. In the immediate future, saved energy is our most promising energy source. Instead of consuming ever more fuel, we can choose to consume the same amount of fuel ever more efficiently.

Making Nuclear
Power Irrelevant

David Morris

While the federal government tries to encourage the growth of nuclear power, many people on their own, in their neighborhoods, or through their city councils or state legislatures are doing whatever they can to avoid the need for nuclear energy. Since it accounts for only 4 percent of our total energy supply and only about 13 percent of our electricity, it is possible that aggressive use of conservation and solar energy could eliminate the need for nuclear power. This is particularly true because electricity is our most expensive form of energy and thus makes solar and conservation efforts the most economical by comparison.

The federal government is sometimes more innovative than the average locality. But the sheer number of smaller governmental units and neighborhoods ensures a great deal of diversity and experimentation. In the most innovative cities it is not uncommon to find that the policies of governments closest to the people are in direct conflict with the policy of the government furthest from the people. For example, the National Energy Plan calls for a 50 percent *increase* in electricity consumption by 1985. But the Davis, California, City Council, after two years of citizen-based planning, established a goal of *reducing* the consumption of electricity by 50 percent in the same time period. And while the federal government promotes increased use of nuclear energy, the City Council of Seattle, after an exhaustive study of the city's future energy needs, rejected a proposed 10 percent interest in a new nuclear power plant and instead established an office of conservation within the municipally owned utility company.

In other respects some of the states are far ahead of the national government. While the federal government has established a goal of 2.5 million solar homes by 1985, California's Governor Jerry Brown has set a goal of 1.5 million solar homes in California alone during the same time. Florida, with 5 percent of the nation's population, plans to meet 20 percent of President Carter's national solar goal by 1985. Hawaii reached its federal 1985 goals by the middle of 1979. Arizona, New Mexico, and California are expected to attain their goals by the early 1980s.

While the federal government fashioned a nominal energy-conservation code for new construction, some states and localities have moved ahead with codes that are more rigorous and climate-specific. There is great potential for energy saving here, because over 20 percent of our energy consumption occurs in our homes. When energy used in commercial buildings is added, the total is over 35 percent.

The Kansas Public Service Commission requires storm windows on new construction. The federal code does not. Seattle's building code goes far beyond that of the federal government. The building code of Davis, California, emphasizes not only building design but land-use planning as well. It permits cottage industry in homes, thus reducing the need to travel. (In Portland, Oregon, a comprehensive energy-conservation study concluded that 5 percent of projected energy use could be saved merely by reviving the neighborhood grocery stores, since people would walk to get a loaf of bread, a bottle of milk, or a pack of cigarettes rather than drive to the shopping center.)

Davis's code requires white rooftops, since air conditioning is a major electricity user in Davis and white reflects heat. Such passive design items (passive means an energy-saving feature that is part of the design of the building and has no moving parts) are a major component in the Davis plan. They are used so widely that in the late afternoon one can see a nearly uniform shadow falling across south-facing windows from overhangs and awnings designed to reduce heat gain from the sun.

Davis is also about to enact an energy code for existing structures. Since our national housing stock changes hands at a rate of less than 3 percent per year, such a provision will have a much more substantial impact in the short term than new building codes. Livermore, California, requires homeowners to meet certain attic insulation standards when they sell their property. Since the average home in Livermore is sold every six years, this ordinance will have a major impact in a short time.

In San Diego, the county commissioners require solar energy to be installed on new construction in unincorporated areas. According to a California state law, contractors who provide a bid on natural-gas swimming pool heaters must at the same time provide cost information on solar swimming pool heaters with a cost comparison over 10 years.

While the federal government has enacted legislation that prohibits utilities from selling, installing, or financing energy conservation and solar energy systems, several cities and states have encouraged the utilities to become directly involved in facilitating the move toward conservation and solar energy.

Oregon private utilities now install conservation measures in customers' homes free of charge upon request. The utilities are repaid, with no interest, when the property is sold or when the owner dies and the property is inherited.

Given an inflation rate of 9 to 10 percent per year, the value of this repayment is so small that the program can be seen as giving away the insulation, storm windows, or other improvements. The state public service commission permits the utilities to put these costs into their rate base, which allows them to earn a profit on the investment. Thus all the customers pay slightly higher utility bills for this service.

In several localities, municipal solar utilities have been designed. Because of opposition by the existing utility to a program whose objective is the reduction of energy consumption based on fossil fuel, these new utilities will probably be established outside the jurisdiction of the existing utilities, even when it is a publicly owned enterprise. In Santa Clara, California, for example, the water department, not the city utility, is leasing solar swimming pool heaters and domestic hot water systems. Since solar heating is a technology that requires, more than anything else,

Tony Auth. Copyright The Philadelphia Inquirer. The Washington Post Writers Group.

a good knowledge of plumbing, the city thought that the water department would be the perfect place for it. In Los Angeles, the mayor's office is investigating the viability of a municipal solar utility. It too will probably be set up outside the Los Angeles Power and Water Department, the largest publicly owned utility in the United States.

The City of Hartford, Connecticut, has gone into direct partnership with small business and neighborhood organizations to encourage conservation and solar energy. The city is establishing a solar manufacturing unit and has already set up a neighborhood energy-audit service that, upon request, examines a residence to see what kind of conservation measures would be most economical.

In Memphis, a unique partnership exists between the South Memphis Economic Development Corporation and its subsidiary, the Sun Belt Solar Corporation, along with Memphis Gas and Electric (a city-owned utility) and the federally owned Tennessee Valley Authority (TVA). In this arrangement the TVA acts as the financing mechanism through a $2 million trust fund. Sun Belt Solar, a minority-owned firm, borrows from that fund to purchase and install domestic hot water systems with no down payment required. Memphis Gas and Electric acts as the billing and bookkeeping agent. The financing has been set up so that customers never pay more for both their electric bill and repayment of the loan than they formerly paid for electricity alone.

Although TVA lends the money in this program under very generous terms (slightly under 4 percent for 20 years), the authority emphasizes that its purpose is not to subsidize solar. Rather, it likes the arrangement because it helps to avoid the need for new power plants. When a home solar system is installed, a large backup tank is also put in. A device installed on the water heater prevents the customer from drawing electricity for backup water heating during times of peak daily electrical use. This causes no inconven-

ience to the customer, because off-peak electricity can easily heat up sufficient water to carry the customer through
peak periods, even during extended periods of cloudiness.
Thus the TVA can defer new and costly power-plant construction, and this financial savings is rebated to the customer through the low-interest, long-term loans.

Sunlight is not the only solar resource localities are pursuing. In Springfield, Vermont, citizens spent three years
analyzing the potential for converting energy from the
stream that runs through the town into hydroelectric
power. When their generating system is complete, Springfield will be selling surplus electricity to the same utility
company it has been paying for power.

A short distance away, the municipally owned utility in
Burlington, Vermont, is using wood from surrounding for-
ests to provide 15 percent of the electrical needs of the
community.

On Block Island, off the coast of Rhode Island, a large
wind turbine provides electricity for the telephone company and many residents, with backup power provided by
the island's old power plant.

It is not only the formal city and state governments that
are seriously pursuing alternatives to fossil-fuel- or nuclear-
based economies. Neighborhoods, antipoverty agencies,
and individual citizens provide their own diverse response
to the energy crisis. In Maui, Hawaii, an area that has the
highest electric rates in the country, an antipoverty organization manufactures solar collectors and supplies them to
low-income residents. In Vermont, the Southeast Vermont
Community Action Agency (SEVCA) has been manufacturing efficient wood stoves for several years. The stoves are
sold to low-income people at cost and to others at a price
that ensures a surplus to SEVCA. In Cranston, Rhode Island,
a community action organization assembles solar collectors
for use in that city.

In San Bernardino, California, the West Side Community Development Corporation (WSCDC), comprising primarily low-income people, has installed the first central solar hot water generation and storage system for residential use. Ten' houses surround the central solar array. WSCDC has trained dozens of low-income residents in the construction skills necessary to find jobs in the flourishing market of solar energy, and the group is currently manufacturing low-cost window-box collectors.

In San Luis, Colorado, elderly and low-income citizens without outside assistance installed 300 greenhouses and hot-air window-box collectors. In Cheyenne, Wyoming, a large greenhouse has been constructed under the supervision of the Domestic Technology Institute for use by elderly and low-income people in raising their own food with enough surplus crop to establish a small cottage industry. In New Mexico, the Solar Sustenance Project sent Bill Yanda to dozens of rural communities to teach residents how to build low-cost greenhouses. A year later he returned to find eight greenhouses built after he left for every one he helped construct. The moral is, once citizens learn the skills of self-reliance, they continue to multiply its benefits.

Other citizen efforts are indirectly related to energy concerns. Citizen associations have helped to enact laws regarding deposits on beverage containers in Delaware, Iowa, Oregon, Vermont, Maine, Michigan, and Connecticut. These laws encourage recycling and returnable beverage containers. Several cities, such as Bowie, Maryland, and Oberlin, Ohio, have enacted similar legislation. More than 200 cities now encourage or require household solid-waste separation and recycling. Since the aluminum industry, for example, uses 4 percent of the total U.S. domestic electrical supply, and a recycled aluminum can requires 96 percent less energy to produce than one made from virgin ore, such recycling can help make nuclear-generated electricity unnecessary. Similarly, recycled copper uses about

85 percent less energy and recycled paper about 30 percent less.

Scores of local groups and individual activists or inventors are making major contributions to new energy systems. The Farallones Institute in Berkeley, California, retrofitted an existing building into the first completely integrated, almost totally self-sufficient house. In an average residential neighborhood, the live-in staff raises enough vegetables, meat, and other foods on their one-fourth-acre site to meet all but their dairy needs. Solar energy provides the hot water they require. A composting toilet takes care of human wastes, and greywater from bathing and washing is recycled into the garden. An attached greenhouse helps heat the building and provides a starting place for plants. A backyard pond produces fish, a side rabbit hutch provides meat, and the curb is covered with alfalfa, which is fed to the rabbits.

In Woods Hole, Massachusetts, the New Alchemy Institute finds itself, after almost a decade, still one of the pioneering research and development organizations on integrated ecosystems. It receives over $100,000 in contributions yearly from its membership to carry on serious, detailed research on microclimates, aquaculture, and alternative energy systems.

Bill Delp and his small company in northern Minnesota are cranking out small-scale hydroelectric systems suited to the small, meandering streams in that area. Charles Schachle, former aeronautical engineer and a resident of Moses Lake, Washington, sells large wind machines at about 10 percent of the cost of comparable demonstration projects funded by the Department of Energy. The list seems endless—Ecotope in Seattle, Center for Neighborhood Technology in Chicago, Center for Community Technology in Madison, Center for Local Self-Reliance in Minneapolis, and Center for Maximum Building Potential in Austin, to name a few.

The smallness of many of these projects is an advantage. As James Madison wrote almost 200 years ago, village-level experimentation is far better than massive pilot projects, because if there is a failure it does not do a great deal of damage, and we can learn from our mistakes. If, however, it is successful, that success can rapidly be transferred and adapted to other localities.

Many of the projects are beginning to develop holistic plans for their turf. In Minneapolis the Center for Local Self-Reliance worked with one neighborhood to analyze energy bills and discovered that only 60 houses were paying more than $75,000 annually in energy payments—enough to provide a solar hot water system for every one of the houses. In the Anacostia neighborhood of Washington, D.C., the Anacostia Energy Alliance has been formed. Three-fourths of the residents in the neighborhood have had their houses audited by neighborhood teams to identify ways to save energy. Residents who have had their houses audited get cards that entitle them to discounts at some local stores. Solar installations are becoming more common.

As innovation bubbles up from below, people begin to consider the potential of energy self-reliance—of cutting loose from increasingly expensive and centralized energy systems. In Franklin County, Massachusetts, not only is energy self-reliance viewed as possible, but a recent study finds that instead of importing energy, the area could be exporting it through wise use of biomass, solar, and hydroelectric sources. In Washington, D.C., a recent study by the Institute for Local Self-Reliance found that more than half a billion dollars is exported from the city to pay for all kinds of energy and that less than 15 cents of the energy dollar ever returns to the city in taxes, dividends, or wages to local residents. The study concluded that D.C. could be almost 50 percent self-reliant with solar energy and waste recovery after economical conservation efforts.

It will take several years to evaluate the impact of these grassroots conservation and solar movements. Yet their exponential growth speaks well for the future. Increasingly, by their personal and collective actions, they are giving the answer to those who ask whether we need nuclear power at all.

The Ultimate in
Preventive Medicine

Helen Caldicott

As a physician, I contend that nuclear technology threatens life on our planet with extinction. Thousands of nuclear
weapons are now being built each year. At this writing, 360
nuclear reactors are in operation in 30 countries around
the world, 70 of them in the United States. Hundreds more
are projected by the end of the century.

Most Americans with whom I've spoken know very little
about the medical hazards posed by nuclear radiation and
seem to have forgotten (or suppressed) the atomic bomb
anxiety that was so prevalent in the 1950s. Today, however, it is of the utmost urgency that we refocus our attention on the problems posed by nuclear technology, for we
have entered and are rapidly passing through a new phase
of the atomic age. Despite the fact that reactor technology
is beset with hazardous shortcomings that threaten the
health and well-being of the nations that employ it, nuclear
power plants are spreading throughout the world. Moreover, by making "peaceful" nuclear technology available
to any nation that purchases a nuclear reactor, we are
inviting other countries to join the international "nuclear
club" militarily, as well as economically. Following India's

lead in 1974, many nations are bound to explode test devices in the coming years. Because of this proliferation of nuclear weapons, the likelihood of nuclear war, the most ominous threat to public health imaginable, becomes greater every day.

Most early developers of nuclear energy explored its potential 40 years ago in an effort to produce bombs that would inflict unprecedented damage. Seven years after the United States detonated two such weapons on the populations of Hiroshima and Nagasaki, the collective guilt generated by the deaths of some 200,000 Japanese civilians prompted the American government to advocate a new policy: the "peaceful use of atomic energy" to produce "safe, clean" electricity, a form of power touted as being "too cheap to meter." Together, industry and government leaders decided that nuclear power would become the energy source of the future. Today, 25 years later, that prospect threatens the well-being of our nation and the world.

I believe it imperative that the American public understand that nuclear power generation is neither safe, nor clean, nor cheap; that new initiatives are urgently required if we are to avoid nuclear catastrophe in a world armed to the teeth with atomic weapons; and that those initiatives must begin with awareness, concern, and action on the part of the individual citizen.

One need not be a scientist or a nuclear engineer to take part in this emerging debate; in fact, an overspecialized approach tends to confuse the issue. The basic questions involved ultimately go beyond the technical problems related to reactor safety and radioactive waste management. Even if the present state of nuclear technology were to be judged failsafe, for example, we must ask ourselves how much faith we would be willing to invest in the infallibility of the human beings who must administer that technology. Granted, we may someday be able to isolate nuclear waste from the environment, how confident can we be in our ability to control the actions of fanatics or criminals? How can we assure the longevity of the social institutions responsible for perpetuating that isolation? And what moral right have we to burden our progeny with this poisonous legacy? Finally, we must confront the philosophical issue at the heart of the crisis: Do we, as a species, possess the wisdom that the intelligent use of nuclear energy demands? If not, are we not courting disaster by continuing to exploit it?

From a purely medical point of view, there really is no controversy: The commercial and military technologies we have developed to release the energy of the nucleus impose unacceptable risks to health and life. As a physician I consider it my responsibility to preserve and further life. Thus, as a doctor, as well as a mother and world citizen, I wish to practice the ultimate form of preventive medicine by ridding the Earth of these technologies that propagate suffering, disease, and death.

Radiation is insidious, because it cannot be detected by the senses. We are not biologically equipped to feel its power, or see, hear, touch, or smell it. Yet gamma radiation can penetrate our bodies if we are exposed to radioactive substances. Beta particles can pass through the skin to damage living cells, although, like alpha particles, which are unable to penetrate this barrier, their most serious and irreparable damage is done when we ingest food or water —or inhale air—contaminated with particles of radioactive matter.

Of all the creatures on Earth, human beings have been found to be one of the most susceptible to the carcinogenic effects of radiation. Because their cells are growing and rapidly dividing, fetuses, infants, and young children are the most sensitive to radiation effects. One of America's most dreaded killer diseases, cancer is like a parasitic organism, often causing slow and painful death. It is estimated that one in three Americans now living will contract the disease at some point. During the 1970s alone, 3.5 million people are expected to die of it.

In addition to giving rise to cancer, radiation also causes genetic mutations, sudden changes in the inheritable characteristics of an organism. A mutation occurs whenever a gene is chemically or structurally changed. Some body cells die or become cancerous when they are mutated; others survive without noticeable changes. A genetically mutated sperm or egg cell may survive free of cancer but can seriously damage the offspring to which it gives rise.

Whether natural or human-made, all radiation is dangerous. There is no "safe" amount of radioactive material or dose of radiation. Why? Because by virtue of the nature of the biological damage done by radiation, it takes only one radioactive atom, one cell, and one gene to initiate the cancer or mutation cycle. Any exposure at all therefore constitutes a serious gamble with the mechanisms of life.

Today almost all geneticists agree that there is no dose of radiation so low that it produces no mutations at all. Thus, even small amounts of background radiation are believed to have genetic effects.

Every medical textbook dealing with the effects of radiation warns that there is no safe level of exposure. Nevertheless, the nuclear industry and government regulatory agencies have established what they claim to be "safe" doses for workers and the general public, drawing support from scientists who believe that there is a threshold below which low doses of ionizing radiation may in fact be harmless. This claim is dangerously misleading and, I believe, incorrect. The International Commission on Radiological Protection (ICRP) originally proposed "allowable" levels of exposure for use by the industry, but not without conceding that these may not be truly safe. Rather, it admittedly accorded priority to the expedient promotion of nuclear power. As the ICRP noted in its 1966 Recommendations, "This limitation necessarily involves a compromise between deleterious effects and social benefits. . . . It is felt that this level provides reasonable latitude for the expansion of atomic energy programs in the foreseeable future. It should be emphasized that the limit may not in fact represent the proper balance between possible harm and probable benefit."

The truth is that we are courting catastrophe. The permissive radiation policy supported by the American government in effect turns us into guinea pigs in an experiment to determine how much radioactive material can be released into the environment before major epidemics of cancer, leukemia, and genetic abnormalities take their toll. The "experts" stand ready to count victims *before* they take remedial action. Meanwhile, the burden remains on the public to prove that the nuclear industry is hazardous, rather than on the industry to prove that it is truly safe.

Today's safety standards have already been shown by several studies to be dangerously high. When investigations of low-doze ionizing radiation revealed that levels of radiation lower than those permitted were causing cancer, government agencies attempted to suppress the findings.

The increasing radiation exposure of workers and the general public by the nuclear industries implies tragedy for many human beings. Increasing numbers of people will have to deal with cancer or, perhaps more painful still, deformed or diseased offspring. It is difficult to predict how many mutated children will be born in the world as a result of nuclear power and weapons production, or what the nature of their defects may be. But it is indisputable that the mutation rate will rise—perhaps far higher than we would care to contemplate. The massive quantities of radiation that would be released in a war fought with nuclear weapons might, over time, cause such great changes in the human gene pool that the following generations might not be recognizable as human beings.

It is important that we keep in mind the fact that the

Trojan Nuclear Plant, Rainier, Washington. Photo by Barry Mitzman.

nuclear industries are relatively young. Nuclear power has been in commercial production in the United States for only 25 years; arms production for 35. Since the latency period of cancer is 12 to 40 years and genetic mutations do not often manifest for generations, we have barely begun to experience the effects radiation can have upon us.

What makes an accident in a nuclear power plant uniquely dangerous is the potential release into the environment of highly poisonous radioactive elements, which can contaminate large areas of land and make them uninhabitable for thousands of years. What makes an accident seem inevitable is the human factor. The most advanced plant is still at the mercy of the fallible human beings who design, build, and operate it. Millions of parts are needed to construct a nuclear reactor, and each must be made, assembled, and operated with little room for error.

In addition to the threat of an accident at a nuclear plant is the problem of radioactive wastes, which are by-products of the nuclear reaction. Can we do anything to protect ourselves and future generations from the lethal legacy of nuclear sewage? At present the answer is no. Technologists have offered a number of ingenious proposals, ranging from solidification of high-level waste in glass containers and burial in salt formations, to lowering waste into ocean trenches, burying it under Antarctic ice, or launching rockets loaded with it into the sun. None of these techniques has been proved to be practical or safe.

Nuclear-industry projections anticipate a total of 152 million gallons of high-level waste by the year 2000. The cost of preparing even our present load of 83 million gallons for geological disposal, however, is currently estimated at $2 to $20 billion. Who will absorb the costs? At present it is not clear. The utilities want the American government (i.e., the taxpayer) to take on the wastes generated by nuclear technology. Since they do not account for the costs of decommissioning plants or waste disposal in their present rates, the utilities can easily present a case for "cheap" energy. Seen in its entirety, however, nuclear energy is far from cheap, and the hidden costs—to our well-being—are enormous.

Succumbing to technical fervor, the U.S. government prematurely committed enormous economic resources, together with political and scientific reputations, to a half-baked technology that is neither cheap, clean, nor safe. The nation's public utilities should not have been permitted to proceed with nuclear energy production until they demonstrated that the public's health could be protected from the carcinogenic and mutagenic effects of its radioactive wastes. This was not done.

Industry engineers and physicists concede that the nuclear waste problem remains to be solved, but in their public pronouncements they urge us to trust them, to have faith in their abilities and in the inevitable advance of technology. I have no confidence in this line of reasoning. It is as if I were to reassure a patient suffering from terminal cancer by saying, "Don't worry, my medical training will enable me to discover a cure."

Nor can technology alone ever provide the answers we seek. For even if unbreakable, corrosion-resistant containers could be designed, any storage site on earth would have to be kept under constant surveillance by incorruptible guards, administered by moral politicians living in a stable, warless society, and left undisturbed by earthquakes, natural disasters, or other acts of God for no less than half a million years—a tall order that science cannot fill.

Because the dramatic decline in the U.S. market for reactors has threatened the financial survival of the nuclear industry, multinational nuclear suppliers, especially Westinghouse and General Electric, are now heavily promoting overseas trade. Their most eager customers are developing nations, such as South Korea, Mexico, Spain, Taiwan, Yugoslavia, and Brazil. Many of these countries lack the capital needed to purchase a nuclear power plant, so American loans are arranged through the Export-Import Bank. They often lack power-transmitting grids to distribute the electricity generated, but the desire to produce electricity is not always their primary motivation: often their ultimate goal in purchasing a nuclear reactor is to gain access to nuclear weapons-grade materials and to join the "nuclear club." India proved this point in 1974; Israel is widely believed to have nuclear capabilities; experts suspect that South Africa may have atomic weapons; and Argentina, Brazil, Pakistan, South Korea, and Taiwan have the potential to develop weapons capabilities in the near future.

This spread of nuclear power plants around the world—and the directly related proliferation of nuclear weapons—seriously threatens global peace and order. At this writing, the United States and the Soviet Union still maintain the balance of power, but the sale of each new reactor tips the scale toward a world of uncontrollable proliferation, in which regional nuclear conflicts could draw the superpowers into all-out nuclear war.

The nuclear industry knows that the reactors it sells produce material for weapons, but its major concern seems to be corporate profit, not morality or human survival. (General Electric is known to have conducted promotional conferences with Egypt and Israel on the same day.)

Nuclear suppliers have, however, voiced concern over the use of reactor by-products for military ends: Corporate representatives from Great Britain, the United States, France, West Germany, Japan, Sweden, and other countries agreed in 1978 that any country buying a nuclear reactor and using its plutonium to manufacture a bomb would receive a "reprimand." Such a scolding would not, of course, preclude further sales to the country at fault.

U.S. government policy toward nonproliferation has been totally inconsistent. Before Jimmy Carter was elected to office, he asserted that nuclear power should be the last resort in America's quest for new sources of energy, and in his inaugural address and first State of the Union message he set the goal of "eliminating nuclear weapons from the face of the Earth." In keeping with this goal, he placed a moratorium on nuclear reprocessing and on the construc-

Central station nuclear power plants in the United States

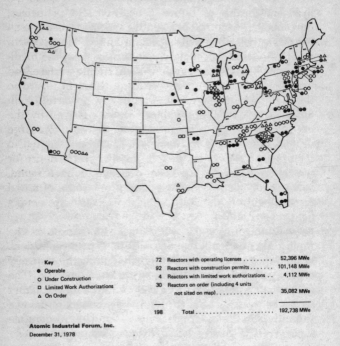

	Key
●	Operable
○	Under Construction
□	Limited Work Authorizations
△	On Order

72	Reactors with operating licenses	52,396 MWe
92	Reactors with construction permits	101,148 MWe
4	Reactors with limited work authorizations . .	4,112 MWe
30	Reactors on order (including 4 units	
	not sited on map).	35,082 MWe
		—————
198	Total	192,738 MWe

Atomic Industrial Forum, Inc.
December 31, 1978

tion of breeder reactors. But in his proposed energy legislation, he shifted his position, calling nuclear power a safe, reliable source of energy and allocating $1.7 billion for its development in his 1978 budget, compared to only $421 million for solar and geothermal research. The Carter administration continues to promote and encourage nuclear reactor sales to developing countries, making reprocessing and breeder moratoriums irrelevant.

My experience in Australia and the United States has taught me many things: first and perhaps most important, that we can no longer afford to entrust our lives, and the lives and health of future generations, to politicians, bureaucrats, "experts," or scientific specialists, because all too often their objectivity is compromised. Most government officials are shockingly uninformed about the medical implications of nuclear power and atomic warfare, and yet they daily make life-and-death decisions in regard to these issues. They are manipulated by powerful, well-financed industrial and military lobbies. Driven by the need for power and ego gratification, they are, to a large degree, desensitized to reality. Their vision is limited to their meager two-to-four-year terms of office; the desire for reelection influences all their decisions. I have found, sadly, that a global view of reality and a sense of moral responsibility for humanity's future are very rare among political figures.

Even after Watergate, Vietnam, and CIA exposés, the American people still seem to trust their political leaders. When I speak at church meetings, describing the enormity of our nuclear madness, people approach me and ask, "The politicians don't know this, do they? Because they wouldn't let any of this happen if they did." I look them in the eye and tell them that their government is totally responsible for organizing this calamity.

But if we can't trust our representatives, whom can we trust? The answer is simple: no one but ourselves. We must educate ourselves about the medical, scientific, and military realities and then move powerfully as individuals accepting full responsibility for preserving our planet for our descendants. Using all our initiative and creativity, we must struggle to convert our democratic system into a society working for life rather than death.

Nuclear power and nuclear war are primarily medical issues. Arguments about profits, jobs, and politics are reduced to irrelevancy when our children are threatened with epidemics of leukemia, cancer, and inherited disease

—or sudden death, in the case of nuclear war. What will be the cost of hospitalizing all the people who contract these terminal diseases? The medical expenses incurred will far outweigh any economic benefits gained by this generation —to the detriment of all those that will come in the future. But of course no human life can be measured in dollars.

It is currently believed that 80 percent of all cancers are caused by environmental factors. By definition, therefore, they are preventable. The U.S. government spends millions of dollars each year funding medical research into the cause and cure of this dreaded disease. At the same time, however, it spends billions funding the weapons and nuclear power industries, which propagate the diseases doctors are struggling to conquer.

America's scientific community has a unique obligation to assess the morality of its work and to assist the public to understand the perils we face: in the event of nuclear holocaust, it is science that will have led us down the road to self-destruction. The first meltdown or other nuclear disaster will disprove, once and for all, the blithe assurances offered by industry experts regarding reactor safety and waste-disposal technology, but only after bringing tragedy into the lives of thousands of human beings.

The industrial age has enthroned science as its new religion; the scientific establishment has, in turn, promised to cure diseases, prolong life, and master the environment. Thus, humanity now believes that it owns the Earth; we have forgotten that we belong to it, and that if we do not obey the natural laws of life and survival, we will all cease to exist.

Unfortunately, my experience has taught me that we cannot rely on our scientists to save us. For one thing, they do not preoccupy themselves with questions of morality. For another, science has become so specialized that even the best scientists are immersed in very narrow areas of research: most lack the time to view the broad results of their endeavor, let alone the willingness to accept responsibility for the awesome destructive capabilities science has developed. When confronted with the realities of our nuclear insanity, they are often either embarrassed or display a cynical fatalism, saying that humanity was not meant to survive anyway. Since about half of America's research scientists and engineers are employed by the military and related industries, they suffer from a profound conflict of interest: many simply prefer not to ponder the consequences of their work too deeply; after all, they have to feed and educate their children—a strange reaction, considering that, as a direct result of their research, their children may not live out their normal life span.

Many scientists *have* stood up to the nuclear establishment, but only at great cost to themselves and their families. Journalist Jack Anderson has observed that these "courageous scientists—Thomas Mancuso, John Gofman, Alice Stewart, George Kneale, Samuel Milham, Arthur Tamplin, Ernest Sternglass, and Irwin Bross—have come under malicious attack reminiscent of the campaign against Hollywood and Broadway liberals during the anticommunist hysteria. We have tried to tell the story of these scientists whose cautious warnings have been assaulted and belittled, whose personal reputations have been besmirched." Why has this happened? Because, Anderson notes,

> The stakes are enormously high. Both the federal government and the nuclear industry are committed to developing nuclear power. Too many unfavorable stories would jeopardize the industry's multibillion-dollar investment. . . . Government officials have also staked their careers on the development of nuclear power. They would look foolish if their massive efforts had to be scrapped because they underestimated the dangers of low-level radiation. Not only would the billions spent on nuclear projects have to be written off, but additional

billions might have to be paid in compensation to those whose health has been impaired.

This intimidation ritual would be less popular with government if more scientists were to accept their moral responsibility to teach the public about the technological dangers to which science has exposed the world.

The huge multinational corporations comprising the

By permission of Bill Sanders. The Milwaukee Journal, 1979. Courtesy of Field Newspaper Syndicate.

THE MILWAUKEE JOURNAL
Field Newspaper Syndicate.

'Frankly, I don't see anything wrong with the watchdog you've got!'

atomic industrial complex influence our elected representatives and appointed officials, through massive federal and state lobbying efforts that reflect their immense power and wealth. Financed by earnings reaped from a variety of technologies that are harmful to the environment and public health, such lobbying can subvert the democratic process by compromising the public interest in favor of the accumulation of ever more profit. As long as our democratic system prevails, however, the general public has an effective recourse in the form of the electoral ballot. The burden lies upon the electorate to become better informed about its genuine long-term interests and then to make its demands known. Public education is the first step, to be followed in part by a grassroots lobbying effort (in the form of a letter-writing campaign and other actions) and personal meetings with elected officials.

We need to educate the politicians in America, by making sure they read about the biological effects of nuclear-fission products and nuclear waste—and I mean *every* federal and state politician in the United States. It is imperative that our political representatives be educated regarding the long-term consequences of their nuclear-related decisions. Because such decisions exert their effects for millennia, they must not be influenced by considerations of short-term self-interest. The public must enable its leaders to understand and reflect popular opinion.

I have testified at hearings held on nuclear power plants in Long Island and Massachusetts. Each time I found myself talking to lawyers employed by the utilities, who know little or nothing about biology. The judges in these hearings are employed by the NRC. Thus, the litigation system surrounding nuclear reactors is often farcical. Subject to obvious conflict of interest, the arbitrators cannot be impartial. I would therefore advise people who are concerned about nuclear power to bypass the legal system and not waste money and energy in an endeavor that will almost always end in defeat. Instead, mobilize in large numbers; march; demonstrate; educate and teach in schools, churches, political meetings—wherever people gather.

The nuclear power plants encroaching on our suburbs make it no longer possible for Americans to bury their heads in the sand and pretend that the fissioned atom does not exist: The radioactive isotopes discharged into the environment by nuclear reactors and nuclear waste are identical to those released when an atomic bomb explodes. Moreover, the spread of commercial nuclear technology to other nations guarantees the uncontrollable proliferation of nuclear weapons. Obviously, something must be done to stop this spread and achieve total nuclear disarmament, but American and Soviet politicians seem unable and/or unwilling to take appropriate action. It is therefore up to us, the people of the world.

Nuclear power plants and military facilities will continue to release radioactive materials into the environment until public pressure becomes great enough to bring such releases to a halt. Because the effects of these materials on us, our children, and our planet will be irreversible, we must take action now. What we have discovered so far should serve as ample warning that our future as a species is imperiled: We are entering a danger zone—an uncharted territory—from which we may never return.

The Roots of the
Antinuclear
Movement

Cathy Wolff

"What's a nuke?" was a common question faced by people with "No Nukes" bumper stickers in 1976. Today they are more likely to receive waves from motorists with similar stickers.

The growth of the United States antinuclear movement has been phenomenal. But it did not happen overnight.

As early as 1959 a citizens' group in Cape Cod successfully fought the dumping of radioactive waste into the ocean. Citizen action also contributed to the cancellation of plans for nuclear reactors in Queens, New York, in 1962; in Bodega Head, California, in 1964; and—through a referendum—in Eugene, Oregon, in 1966.

Nuclear power technology grew out of the development of the atomic bomb. And much of the opposition to nuclear power is rooted in the 1950s movement against atomic weapons. The two movements are seen by many people as being inseparable and clearly to have nurtured each other.

Antinuclear rally in Washington D.C., May 6, 1979. Photo by Frank Johnston, The Washington Post.

For example, demonstrations against the Trident nuclear submarine in recent years drew hundreds of people, many of them veterans of anti–nuclear-plant actions. The Mobilization for Survival, a national organization formed in 1977 with a primary focus on halting nuclear weapons, soon broadened its emphasis to include nuclear reactors and has done a lot to help demonstrate the connections between the two.

The historical roots of the shape and power of the antinuclear movement can also be traced to the Vietnam War and Watergate—two recent events that made many Americans skeptical of government promises.

The environmental movement helped create a distrust of dependence on centralized technology and fostered an awareness of the Earth's vulnerability.

Nuclear opposition lends itself to wide-based support because it's a broad issue. Some facet of it can be found in everyone's backyard. Unlike opposition to the Vietnam

War, the antinuclear movement was not born on college campuses. Native Americans are opposing uranium mining on their reservations. Farmers in upstate New York and Minnesota are fighting the erection of dangerous high-voltage power lines. Radioactive waste dumps—existing or planned—have sparked active citizen opposition in many states. The transportation of nuclear materials is rousing concern among airline pilots, truck drivers, railroad engineers, and people who live near the atomic shipment routes. Workers in nuclear facilities are growing increasingly aware of radiation dangers.

And women, especially mothers, are particularly worried about radiation-caused birth defects in future generations. Several polls, including those by the nuclear industry, have shown women to be much more likely to oppose nuclear plants than men. In fact, women have often taken the lead in grassroots antinuclear organizing and in legal interventions.

Intervention was the first tactic many nuclear opponents turned to in the mid-1960s. It involves participating in the regulatory procedures that are required before a nuclear plant can be built.

Intervention led to many important safety advances in nuclear plants. The intervention against the Vermont Yankee plant in Vernon, Vermont, forced the company to use cooling towers to minimize thermal pollution of the Connecticut River. One woman's intervention in hearings for the Seabrook, New Hampshire, reactors led to a major strengthening of the plant's design so it could better withstand earthquakes.

The intervention process has helped to slow the spread of nuclear plants, to gather immense amounts of data, to monitor the industry, and to document the pronuclear bias of regulatory agencies. Perhaps most important, it has laid the groundwork for the decentralized, grassroots, action-oriented emphasis of the movement, because many people became frustrated with drawn-out proceedings that often ended in defeat.

There have been more intervention success stories in connection with nuclear-related electric rate-hike cases than with the nuclear plants themselves.

In Connecticut in 1977, for instance, intervention led to denial of a $90 million rate-hike request by Northeast Utilities. The company also was ordered to withdraw its partial ownership of the Seabrook plant and postpone plans for two nuclear reactors in Montague, Massachusetts.

But legal intervention is time-consuming, complex, and expensive and requires legal expertise. Lengthy technical hearings on thermal pollution or the height of cooling towers fail to grab headlines or spark widespread public interest.

Intervention, by its nature, is myopic; most intervening groups are forced to challenge specific environmental impacts of particular plants rather than address the entire nuclear question. Finally, it is an exasperating process, usually undertaken before a stacked jury of nuclear advocates.

It was this frustration that led to new tactics—tactics designed to bring the question of nuclear power directly to the people.

Statewide referenda on nuclear plant construction in 1972 and 1976 were defeated in six states. But in 1976 Missouri voters outlawed a utility practice of charging consumers for the financing costs of nuclear plants under construction—another example of economic arguments having more appeal than safety or moral questions.

In 1978 voters in three states—Montana, Oregon, and Hawaii—enacted measures that put severe restrictions on nuclear development.

Although initiative drives have the advantage of drawing wider public interest to the issue and giving a specific focus for more broad-based organizing, they are slow, expensive, and frustrating, especially since the nuclear industry consistently outspends nuclear opponents by ten to one or more.

Some believe more dramatic action is needed to draw attention to the nuclear power issue.

On February 22, 1974, Sam Lovejoy toppled part of a weather-monitoring tower Northeast Utilities had erected on the proposed site of a nuclear plant near the farm where Lovejoy lived in Montague, Massachusetts.

Lovejoy used his trial as a forum to present the dangers of nuclear power and a case for nonviolent civil disobedience.

Charges were dismissed because of a legal technicality, but the trial helped make nuclear power a major issue in western Massachusetts, accelerating local organizing.

Antinuclear activists, including Lovejoy, ran for local office, gathered signatures on antinuclear referenda petitions, and held public meetings, rallies, and teach-ins.

Perhaps one of the most exciting educational projects focused on showing people the alternatives to a nuclear-powered future. The Toward Tomorrow Fair, organized at the University of Massachusetts in Amherst, included hundreds of exhibits on alternative energy technologies, drawing thousands of people from throughout New England. It has become an annual summer event.

In the winter of 1976, farmer Ron Rieck climbed to the top of a weather-monitoring tower in Seabrook, New Hampshire, and stayed for 36 hours in protest of plans for a nuclear plant there.

But it was the Clamshell Alliance that popularized mass civil disobedience in the United States antinuclear movement. The loose-knit New England–based organization was formed in July 1976, after the Nuclear Regulatory Commission issued a construction permit for a nuclear power plant at Seabrook, New Hampshire, ignoring years of intervention by several citizens' groups.

The Clamshell Alliance, inspired by the takeover of a

nuclear site by thousands of Europeans in Whyl, West Germany, decided to occupy Seabrook.

The first Seabrook occupation deliberately involved only 18 people, all New Hampshire residents who wanted to show support for a Seabrook town vote against the plant. That was August 1, 1976. On August 22, 1976, people from all over New England walked onto the site—180 strong. Less than a year later, on April 30, 1977, the number had grown to 2000, and they came from across the United States. More than 20,000 people came to the site June 25, 1978, for a legal demonstration sponsored by the Clamshell Alliance.

It was the 1977 arrest and two-week jailing of more than 1400 people that grabbed international headlines and spurred the development of dozens of similar alliances around the country—the Abalone in California, the Sunflower in Kansas, the Keystone in Pennsylvania, the Northern Sun in Minnesota, the Crabshell in Oregon, and many others.

Most of these groups adopted Clamshell's consensus decision-making process, inspired by the Friends (Quakers); its nonhierarchical, open-membership style of organization, and its focus on well-organized, nonviolent site occupations.

And like the Clamshell Alliance, they soon had to deal with problems such as trying to reach consensus among hundreds of people with a wide range of motives, experience, politics, and style; trying through bake sales, loans, and sale of resource materials to keep themselves financially alive; and constantly reevaluating tactics and strategy.

The Clamshell Alliance and other groups consistently emphasize the importance of local and regional work, keeping in touch with other regions' activities through a wide exchange of newsletters and constantly putting people in touch with other activists in their areas.

Education work is ongoing. The Clamshell Alliance organized a conference on the morality of nuclear power for New Hampshire clergy, a conference on jobs and energy for labor leaders, and a speakers' bureau for public schools.

Two films produced by Green Mountain Post Films of Turners Falls, Massachusetts, became mainstay organizing tools for antinuclear groups around the nation—*Lovejoy's Nuclear War*, about Sam Lovejoy and nuclear power, and *The Last Resort*, about the growth of the Clamshell Alliance.

The films are only part of the growing cultural aspect of nuclear opposition. Performers, writers, and artists have been attracted to the movement and have created many songs, poems, plays, novels, and posters to dramatize the issue. As the movement swelled, nationally known performers lent their talents. Jackson Browne, Arlo Guthrie, John Hall, Pete Seeger, and others have held many benefit concerts to raise money for the movement.

The China Syndrome, a dramatic Hollywood movie about a nuclear accident starring antiwar and antinuclear activist Jane Fonda, gave the movement an unexpected push. Ironically, the movie was released within a few days of the near-meltdown at Three Mile Island.

But even with *The China Syndrome*, the Three Mile Island accident probably would have failed to gain front-page international attention if the media had not been primed for three years by high profile antinuclear actions. Earlier accidents, potentially as serious as Three Mile Island, had gone virtually unnoticed by the media in the past.

Following the accident, nuclear power became a legitimate issue for many liberal politicians, the press, and others. Suddenly, town officials near nuclear plants were demanding evacuation plans. Nuclear moratorium proposals proliferated, and many people were taking a second look at antinuclear arguments. More than 100,000 people went to Washington, D.C., on May 6, 1979, to hear these arguments. It was the largest antinuclear gathering in U.S. history.

The strength of the antinuclear movement probably will continue to come from grassroots, local work. An increasing number of national and Washington, D.C.–based organizations are contributing tremendous resources to the movement. But it is through the local organizers—talking with their neighbors, confronting local nuclear threats, and building local movements—that these resources have an impact.

The nuclear industry is not dead. It will take years of work to shut down uranium mines, dismantle existing nuclear plants, provide the funding for safe energy alternatives, and end the nuclear threat.

Even then—when nukes finally are halted in the United States—nuclear exports, particularly to Third World nations, may continue. It is imperative that the same energy go into fighting nuclear plants everywhere—not just in our backyards.

Safe Energy Begins at Home

*Harvey Wasserman
and Lee Stephenson*

Where were you when you first heard about the Three Mile Island (TMI) nuclear plant accident?

Were you close enough to the southeastern Pennsylvania site to need to worry about the threat of radiation? Although government and company officials never told us so, more than 100 million people within several hundred miles were in the danger zone.

Did you think about what it would be like to flee your home from a deadly menace you could not see, hear, feel, or smell but which threatened to plague your family for generations with cancer and birth defects? Could you have prepared yourself for the fact that in the event of a major accident that heavily contaminated your community, you might never be able to return?

For millions of people on the East Coast such questions did come to mind on March 28, 1979, and in the following days. Perhaps those of us who didn't think about it, should.

It may be only luck that prevented a meltdown at TMI. The operators of the plant were kept guessing about conditions in the core, and they often guessed wrong. They were groping in the dark. A meltdown or explosion could have ruptured the containment vessel. Depending on wind conditions, that disaster could have sent a radioactive cloud over Philadelphia, Baltimore, Washington, D.C., New York City, or even Boston. There is very little chance that any of these urban areas with their millions of inhabitants could have been successfully evacuated in the time between the occurrence of a major radioactive release and the poison's arrival at the city. The warning time is wholly dependent on the weather and the competence of the officials monitoring the accident.

Think about this for a minute. In the event of a serious nuclear accident, we're talking about thousands of deaths outright and additional thousands of people surviving with cancer or passing on genetic defects that will plague generations; the rendering of a land mass the size of a large state uninhabitable for thousands of years; and contamination of water and air that could have global repercussions. We're discussing a unique and terrifying threat.

It is difficult to accept even now. But in the final days of March we really were on the brink of a catastrophe that defies the imagination. Even though catastrophe was averted, radiation did leak in dangerous quantities from the plant. For those who were downwind nearby, only time will reveal the measure of the disaster that did occur. Secretary of Health, Education, and Welfare Joseph Califano has estimated that one to ten people who were within a 50-mile radius of TMI will die from cancer. Clearly, others will contract the disease but survive and many will pass on genetic defects to their children and grandchildren and great-grandchildren and beyond.

The TMI accident was neither the first nor the worst that the nuclear age has given us. You have already read in this book about an accident at the Tennessee Valley Authority's Browns Ferry reactor in Alabama, which could have been just as serious as TMI.

There were others. In 1958 a partial meltdown at a reactor in Windscale, England, contaminated a vast area of farmland, led to the confiscation and slaughter of thousands of farm animals, prompted the dumping of large quantities of milk into the Irish Sea, and caused a marked increase in radiation levels over London, 300 miles away.

In 1966 the Fermi One reactor at Monroe, Michigan, stood on the brink of catastrophe for a full month. Because it was an experimental breeder reactor and used highly volatile liquid sodium for coolant, the odds on it causing a major explosion may have been far greater than those at Three Mile Island. Throughout October 1966, worried state and federal officials contemplated evacuating Detroit. Yet the accident entirely escaped media attention at the time.

Serious accidents have occurred at several experimental reactors in Canada and the United States. And a 1950s radiation disaster of unknown origin in the Soviet Union rendered a huge land mass uninhabitable and may have killed hundreds of people. Only sketchy details about that accident are available in the West.

If you place yourself in the shoes of future generations looking back on this era, you're likely to find yourself wondering, "What were they thinking of when they let nuclear power happen?" It will probably be a reaction similar to the sad wonder we experience when we reflect on the most destructive and nonsensical of humanity's failings through the ages. Why have we allowed a misdirected government and private interests, whose primary goal is profit, to saddle us with this deadly technology that we could easily avoid? Why haven't more people among us stood up and said no?

Perhaps the saddest comment of all is a crass attempt to exploit public fear generated by the TMI accident. Within weeks of the radiation leaks, agents of the Mutual of Omaha insurance company started going door to door in the towns near the accident site selling cancer insurance. In the first week of their campaign they easily sold more than 100 policies to residents nervous about their future health. It seems incredible that we can accept the risks of nuclear power and then, when an accident does occur, simply buy a little bit of extra insurance. Are we so helpless that we'll settle for a little paper protection rather than demand that the deadly industry be shut down?

As we have seen in this book, the U.S. government is largely to blame for leading us down the nuclear path. Government officials who had been responsible for the atomic bomb the United States dropped on Japan desperately wanted to find peaceful uses for nuclear technology —both because they wanted to ease their consciences and because there were millions of dollars invested in it. It took quite a few years and a lot of incentives to hook the nation's electric utilities to the nuclear bandwagon. They drove a hard bargain.

Among other incentives, the utilities wanted a limit on their liability in the event of a nuclear accident. Congress gave it to them in 1957 with the passage of the Price-Anderson Act. As Senator Mike Gravel pointed out in an earler article, the $560 million liability limit set by the act is far less than the damage a major accident would cause. In addition, the 20-year limitation on claims is absurd, because many cancer cases and genetic defects will not show up until long after that time has passed.

Once the utilities did buy in, they and the companies that design, build, and finance nuclear plants became a potent force to steer the nation ever more deeply into a nuclear quagmire from which they were in a position to reap enormous profit.

The government and the private nuclear industry have also worked together over the years to mystify and keep secret the facts about nuclear power that would have allowed the public to understand and make decisions about

Tony Auth. © 1979, The Philadelphia Inquirer. The Washington Post Writers Group.

its use. They avoided research projects that might come up with unwanted answers. For example, the government and industry have never systematically sought to determine the health effects of low-level radiation to which they are determined to expose us. For years both the government and industry have stood by silently while workers in uranium mines—many of them Native Americans—contracted cancer at rates far higher than those for the rest of the population.

And when ongoing government studies failed to support the industry, they were often canceled and their authors harassed. As we read earlier in this book, this was the case for Dr. Thomas Mancuso, whose long-range study of the health of workers at nuclear facilities was canceled by federal energy officials after he refused to misuse his data in a government attempt to downplay radiation dangers.

Is it feasible to stop building more nuclear plants? Can

we shut down the ones we now have? Practically speaking, this is the bottom line of the nuclear power debate.

The United States currently has more electric power plants than it needs—a surplus generating capacity of about 37 percent. This figure is based on the peak-use periods of the summer, when the nation as a whole uses more electricity than at any other time. Utilities need up to a 20 percent surplus margin to allow for breakdowns and routine equipment maintenance. That leaves a 17 percent excess.

Seventy nuclear plants now operate in the United States, representing 10 percent of our electrical generating capacity.* This means that we could theoretically shut down all of our nuclear plants right now and still have a 7 percent excess reserve margin—more than enough backup power to carry on business as usual.

However, this step might cause electric bills in many parts of the country to increase. Fossil-fuel power plants (usually coal), which are now part of that excess capacity, might need to be brought back into full use. Their fuel costs are higher than those of nuclear plants, but as Charles Komanoff explained earlier in this book, this doesn't mean that coal plant electricity is more expensive than nuclear. In fact, when construction costs are included, it is less expensive.

If all nuclear plants were closed, electricity bills might rise an average of 6 to 8 percent nationwide. Considering the health and safety dangers of nuclear power, that seems to be a bargain. But the increases would not be divided equally across the nation. Areas that already use fossil-fuel plants almost exclusively, such as the Plains and Rocky Mountain states, would suffer few increases. But regions that rely heavily on nuclear power, such as the Chicago

*Nuclear power plants supply 13 percent of the electricity we actually *use*, as is noted elsewhere in this book.

area, New England, and the Carolinas, might face doubled electric bills.

There also might be some difficulty in distributing the electricity generated by fossil fuel to the places that need it the most. An electricity transmission "grid" connects many electric utilities and makes it possible for them to buy and sell power to one another. But the grid does not now connect all points in the system, so it can't always distribute power from the utility that has it to the one that needs it.

Another consequence of shutting down all existing nuclear plants and halting the 80 under construction would be the need to begin building new coal-fired power plants. Assuming that U.S. demand for electricity continues to grow at its current annual rate of about 3.5 percent, the 7 percent excess reserve margin would be exhausted in about two years if all nuclear plants were closed.

This need for new coal-fired electricity could, of course, be greatly reduced or even eliminated by a reasonable commitment to halting waste of electricity. This might be difficult for some fast-growing areas of the nation that are counting on nuclear plants now under construction to supply their expanding electricity needs. But many of these booming areas are in the Sun Belt, where solar energy is a particularly attractive—and obvious—alternative. By moving toward solar energy and conservation, these growing areas could virtually eliminate the need for new nuclear power plants.

We believe the United States should shut down all its nuclear plants. A few of the plants in particularly atom-dependent regions may be needed until substitute plants are built and effective conservation measures take hold. These remaining nuclear plants could be phased out over time.

We think it is clear that we can shut down most nuclear plants without having to make any immediate changes in the way we live. Our need to build more coal-fired plants to replace nuclear facilities is directly related to our willingness to take sensible steps to waste less energy—and to insist that our government does the same.

The price of energy will go up whether or not we shut down nuclear energy. The rising costs of building and operating nuclear power plants will far outstrip the price of increasingly expensive fossil-fuel-generated electricity. Near-total reliance on fossil fuels might raise our electric bills in the short term, but those increases would certainly be less than if we continued to build nuclear plants. We can lean on fossil fuels as we develop solar technologies and conservation efforts. In the long term, as the many forms of solar power become the basis of our energy system, we can look forward to cheaper and safer energy than if we follow any path advocated by the energy monopolies. The sooner we get started, the sooner we can reap these benefits and rid ourselves of the risks of nuclear power.

It is time for the American public to exert control over its energy system. We were never asked if we wanted nuclear power, and when citizens have been given a chance to vote on the question they have often rejected it. In other cases, well-financed industry campaigns have convinced people to vote for nuclear power by using lies about its ability to provide safer, more reliable energy and more jobs.

As Denis Hayes has pointed out, the United States wastes more than 50 percent of the energy it uses—12 times what we get from atomic power. Simple improvements, such as energy-conscious design of buildings, upgrading mass transit systems, rejuvenating the railroads, recycling trash, and refurbishing home heating, cooling, and insulation systems can save us amounts of energy dwarfing what we now get from nuclear.

Promising "new" sources (they're actually the oldest we know) such as capturing the sun's rays for heating and cooling our homes, burning wood, fermenting organic products into fuel alcohol, revitalizing our hydroelectric dams, and revamping our windmill industry can provide yet again more power than nuclear plants. Other alternatives —such as photovoltaic cells that turn sunlight directly into electricity—are available but are still undergoing development to reduce their cost to competitive levels.

These "renewable" technologies are generally more ecologically sound than our current sources of energy, particularly nuclear power. According to the President's Council on Environmental Quality, the United States could get 25 percent of its energy from renewable sources —more than six times the proportion we now get from nuclear—by the year 2000. The cost of such a program, according to the Solar Lobby's *Blueprint for a Solar America*, would be roughly $2.5 billion per year, the same as the price tag on an average new twin nuclear-reactor complex.

Such a program, if carried into the next century, could yield us a nearly totally solarized energy system by the year 2050. That would drastically reduce our fuel costs and would free us from some of our biggest sources of both pollution and inflation. It would guarantee us energy independence.

And turning to alternative sources would revolutionize the job market. Renewable energy technologies are labor-intensive, meaning they rely more on human work than on invested capital. The bulk of the money that goes into building atomic power plants pays for exotic materials, complex machinery, and expensive technical expertise. Proportionately little of it goes for labor, and the jobs that are created are highly specialized. A $2 billion nuclear reactor provides permanent jobs for only 200 people.

Solar power, recycling, and increased energy efficiency rely far more heavily on human labor, creating jobs that are accessible to working people with a far wider range of skills. Up to seven times as many jobs may be created per

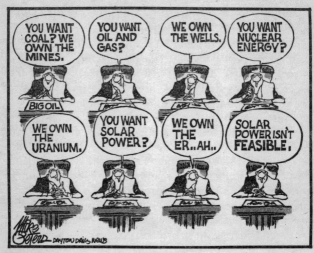

Mike Peters. Copyright Dayton Daily News.

dollar spent on renewable technologies as are created by a dollar spent on nuclear. "Today," says United Auto Workers President Douglas Fraser, in an earlier article in this book, "when our nation needs both safe energy and decent jobs, solar power can provide both. . . . It would move this country closer to its crucial goals of energy independence and full employment."

The political ramifications of such a move would also be enormous. The major social attribute of solar technology is that it is by nature best suited for small-scale, decentralized applications. With simple material inputs, relatively elementary labor requirements, and universally available fuel, solar power is the ultimate form of energy for a democratic society. It can be managed and controlled by individual homeowners, neighborhoods, and small communities. It can be—and is beginning to be—used by consumers everywhere to cut their dependence on utility companies and the energy monopolies.

But the control of nuclear power is out of our hands. It is the exclusive property of a financial and technical elite and is best designed, ultimately, for keeping power sources centralized—perhaps the only function it serves with any real reliability.

We've seen in this book that nuclear power is neither safe nor economical. We've seen that major accidents and near-misses—with their attendant destruction, disease, and death—have taken place and that more such occurrences are inevitable. In addition, long-term isolation of wastes is at present a technical fantasy. Safe burial sites have not been found, and the idea of burying wastes may itself be unworkable. In addition, it is impossible to handle and transport highly radioactive substances without having some escape into the environment.

We also know that the technology for decommissioning atomic reactors is not perfected and that the sites themselves may have to be guarded for thousands of years after the brief 30-to-40 year life span of the plants.

It is clear that nuclear power cannot be separated from the proliferation of nuclear weapons because by operating the power plants the materials for atomic weapons are inevitably produced. This makes the spread of nuclear weapons—both to nations and to terrorists—impossible to control.

We've seen that the nuclear industry and the government are so sensitive about criticism of their policies that they have resorted to surveillance, harassment, and perhaps, in the case of Karen Silkwood, the murder of critics.

We put up with all this for the measly 4 percent contribution that nuclear power makes to America's energy supply. And the government uses a fourth of that just to operate the enrichment plants to prepare uranium fuel for reactors.

"SAY, ISN'T THAT THE SAME BUNCH WHO ONCE RAN THE PRESIDENT OUT OF OFFICE?"

Marlette. Copyright © 1979, The Charlotte Observer. King Features Syndicate.

The Carter administration entered the White House having made a pledge that nuclear power should be "a last resort." However, the administration's performance has been far more pro-nuclear and less pro-solar than that pledge indicated. Congress has also consistently backed the nuclear industry. In the long run, it seems obvious that only a powerful, well-organized citizens' movement can make the kind of lasting changes in our energy system that we need.

That means that it's up to us, which is just as well, since we should have been involved in the decisions all along. However, now it will be much harder for us to exert our collective opinion, because a powerful structure has evolved to promote and protect the nuclear industry.

For years, a small number of people have worked to organize citizens against the nuclear menace. It started with citizens who wanted to stop the spread of nuclear weapons after World War II, and a second front developed with the coming of nuclear power. Early criticis spent much of their time involved with official licensing proceedings or lawsuits. As more people became involved, individuals and groups began to hold public meetings and demonstrations to voice their opposition. By 1978, even before the Three Mile Island accident, the antinuclear movement had become a full-scale social upheaval, reminiscent of the early days of the Vietnam War. More than 20,000 people had already rallied at a nuclear construction site in the small, New Hampshire town of Seabrook. And thousands more around the country had marched against reactors in their own regions.

However, the electric utilities and the other giant oil, natural-gas, coal, construction, and financial companies that make up the energy industry have spent many millions of dollars to sell their vision of the future, and many of us, whether or not we care to admit it, have been influenced by their appeals. What their slick campaigns say, to put it bluntly, is that unless we follow their blueprint for the future, we'll freeze in the dark. They want us to use more and more energy from the costly and dangerous sources they monopolize. The policies they advocate require unprecedented outlays of capital. But because these proposals are familiar they seem less threatening than a major commitment to increased energy efficiency and solar technologies.

Utilities and the energy industry don't want us to conserve energy unless they can keep charging us higher prices to make up for quantities we don't buy. The companies usually imply that conservation means higher unemployment and a weakened economy. Energy conservation may make sense, but the business of utilities and energy companies isn't to develop a rational energy policy. Their business is to make money. The two goals often aren't compatible.

The energy industry won't sit by passively if we attempt

to turn away from nuclear power. They've invested too much capital to give up without a big fight. They'll do everything they can to convince us all that we're alarmists. But as threatening as it is to make a break from the status quo, we all sense that the industry is not working with our best interests in mind and that the government is too heavily influenced by industry to watch out for us.

Here's what we can do. Talking to friends and neighbors about energy problems is an important beginning. When we share our concerns with others, it gives them the courage to speak out too. Groups working in your area can help provide information and ideas on how to take action. See the resources listed in the back of this book to help you find them.

Letter-writing campaigns can be very effective. In addition to your federal and state legislators, write to your mayor, governor, and state public service commission (sometimes called public utility commission). This often obscure agency of your state government is responsible for regulating electric utilities and determining how much you pay for power. Tell them how you feel about picking up the tab for cost overruns in nuclear power plant construction. And tell them your feelings about paying for nuclear accidents (like the customers of TMI will do if Metropolitan Edison has its way). Tell your state commission you want them to stop allowing utilities to use the financing gimmicks that help them build nuclear plants. A feature called CWIP (construction work in progress), which is used in many states, lets utilities charge us for power plants that are still under construction. And through so-called adjustment clauses, utilities are often permitted to charge us automatically for the extra costs of buying electricity from other utilities when a nuclear plant breaks down or is closed for safety reasons.

Some ratepayers throughout America have begun withholding that portion of their bill destined to pay for atomic power. They have begun attending the hearings of their public utilities commissions, demanding that atomic reactors be taken out of the rate base, and that rate hikes slated to go for new reactors be slashed and canceled. Such citizen action has had a genuine effect, leading directly to the cancellation of numerous atomic projects.

Do some research to learn where your area's energy comes from. Where is the closest nuclear plant to your home? Are you paying for it in your electric bill? If there is one within several hundred miles, you may want to inquire into its safety record.

Ask your local health department if it takes regular measurements of radiation levels in the air and water in your area. How are radioactive materials shipped to and from nuclear facilities in your region? They may go down the main street of your town. Since nuclear cargoes are moved long distances, you may have shipments passing through your area even if the nearest nuclear facility is 500 or more miles away.

Let's stop waiting for the government to decide for us to increase the efficiency of our energy system. We can start on our own. Conservation is the cheapest, least-polluting energy alternative we have. All we need to do is organize our efforts and convince our neighbors to join us. Home insulation, car pools, mass transit, recycling, and buying more efficient appliances are just a few of the possibilities. We can start to reap the multiple benefits of saving some of that 50 percent of energy that we now waste and eliminate the need for nuclear power while we're at it.

We can start paying more attention to opportunities to use solar energy—whether it's a local ordinance that encourages passive solar devices such as awnings and light-colored roofing, or commitments to solar-warmed water and home heating.

Some of these alternatives may require some investment of time and money. The system isn't geared to support such alternatives, so you often have to break your own ground. But there are lots of other people who are looking for alternatives too.

And take a look at how much is being spent to build the nearest nuclear plant and how much we're handing over to the major oil and coal companies. That's our money, and where it goes for energy is a political decision. There is no law that says our energy dollars have to continue to support the kinds of power we don't want. The Seabrook nuclear project now represents some $20,000 per household in New Hampshire, more than enough to solarize the whole state. The billion dollars spent to build Three Mile Island could have gone a long way toward cutting traditional energy demand and installing renewable technologies in central Pennsylvania and New Jersey. Instead of cheap, clean, safe power, we got a health and financial disaster.

Public pressure should also be applied to our congressional representatives to repeal the Price-Anderson Act, which throws the burden of nuclear insurance onto the taxpayer. If atomic power is as safe as the industry says it is, then the industry should cover the full cost of any accident, without the benefit of public charity. Congress passed that law, and Congress—our representatives—can repeal it.

Congress can also end the enormous subsidies that continue to go to the atomic industry in the form of funds for research and development as well as huge subsidies for uranium enrichment. The companies that developed atomic power for private profit should take their own private losses, without bail-outs from the public treasury.

The continued push for nuclear power, in spite of continued public opposition, has driven thousands of citizens to be arrested in demonstrations to stop it. If the program continues, we can only expect many many thousands more arrests, with all the social polarization and human cost that entails.

If the nuclear power continues, we can expect higher

electric bills, more inflation in the energy sector, more Three Mile Islands—and worse.

The time to end this tragic, failed experiment is now. Individual reactors have been stopped all over the world; the industry as a whole can also be stopped.

We can get all the power we need from the sun, from recycling, and from trimming the inefficiencies of our current energy system—and we can do it all cheaper, more reliably and more democratically, creating more jobs as we go, than through any method the energy cartels can ever offer.

We can immediately shut down nearly all our existing nuclear plants and phase out the remainder as we increase our commitment to energy efficiency and solar power. The result would be a cheaper, safer, more reliable energy system more in tune with a democratic way of life and the source of more jobs than the highly capital-intensive schemes offered by the energy industry.

But make no mistake about it, there are major barriers in our way. No president, no single act of Congress, no single demonstration or letter can alone move those energy monopolies.

What can do it is individual citizens, neighborhoods, and communities deciding for themselves what kind of power should be generated and how.

The money is there if we take the political power to see it used for the common good rather than for corporate profit.

The technology is there if we can gather the human energy to organize and grab hold of it.

And the person who will decide the outcome, neighbor, is you.

© 1979 by Laurie E. Usher.

ABOUT THE AUTHORS

Dr. Helen Caldicott was born and educated in Australia, where she made a major contribution to the movements to ban the mining of uranium and halt French atmospheric testing of nuclear weapons in the South Pacific. Now a resident of the United States, she is a pediatrician with the Children's Hospital Medical Center in Boston and the author of *Nuclear Madness: What You Can Do.*

David Dinsmore Comey was formerly executive director of the Chicago-based Citizens for a Better Environment. He was killed in an automobile crash on January 5, 1979. He is remembered as one of the most effective and persistent critics of nuclear power.

Jerry Elmer is a field secretary for the American Friends Service Committee, a Quaker service and educational organization.

Andy Feeney is editor of *Power Line,* the journal of the Environmental Action Foundation's Utility Project. He was formerly communications coordinator for the Citizens Action Coalition of Indiana and was the recipient of an Environmental Protection Agency award in 1974 for environmental reporting in Indiana.

Douglas Fraser is president of the International Union, United Automobile, Aerospace, and Agricultural Implement Workers of America (UAW).

Mike Gravel is a Democratic U.S. Senator from Alaska. He has introduced numerous bills that would place a moratorium on nuclear power plants and promote solar energy technologies.

Denis Hayes is the director of the Department of Energy's Solar Research Institute. He was previously a senior researcher with the Washington, D.C.–based Worldwatch Institute and the chairman of the Board of Directors of the Center for Renewable Resources and the Solar Lobby. He is author of *Rays of Hope: The Transition to a Post-Petroleum World.*

Mark Hertsgaard works for the Institute of Policy Studies in Washington, D.C., and is writing a book about the American nuclear power industry to be published in the spring of 1980. A summary

315

of that work will appear in the Institute's forthcoming study of the transnational nuclear power industry.

Charles Komanoff is an energy consultant and adviser to the General Accounting Office of Congress, the New York State Consumer Protection Board, and the New Jersey Public Advocate, among other government agencies. He is the author of *The Price of Power: Electric Utilities and the Environment* and *Power Plant Performance: Nuclear and Coal Capacity Factors and Economics.* The latter book and his subsequent studies have been instrumental in exposing the true economic costs of nuclear power.

Sam Love was one of the founders of Environmental Action and is now an energy consultant and writer, as well as a lecturer on the future.

Richard E. Morgan is the research coordinator of the Environmental Action Foundation's Utility Project and has worked for citizens' groups in Ohio and New Mexico. He is the author of *Nuclear Power: The Bargain We Can't Afford* and a coauthor of *How to Challenge Your Local Electric Utility* and *Taking Charge: A New Look at Public Power.* He is currently writing a book on utility rate structures.

David Morris is an urban planner and a director of the Institute for Local Self-Reliance in Washington, D.C. He is coauthor of *Neighborhood Power,* a regular columnist for *Solar Age,* and a director of the D.C. Energy Self-Reliance project. He has been a consultant to neighborhood organizations and municipal, state, and federal governments.

William T. Reynolds is the coordinator of the Nuclear Cargo Transportation Project, which conducts research and works to oppose the transport of nuclear weapons and nuclear wastes. The organization is a project of the American Friends Service Committee. He also has worked as a consultant on coastal environmental affairs.

Lorna Salzman lives in New York City and has for many years been the mid-Atlantic representative for Friends of the Earth.

Lee Stephenson is a writer and editor on energy and ecology topics. He is the author of the *Energy Workbook for Parks* and coeditor of the *Guide to Corporations: A Social Perspective.*

Kitty Tucker is the president of Supporters of Silkwood, as well as the Health and Energy Learning Project (HELP), and is a graduate of Antioch Law School.

Donna Warnock works with Feminist Resources on Energy and Ecology and the Syracuse Peace Council. She is the author of *Nuclear Power and Civil Liberties: Can We Have Both?*

Harvey Wasserman is a long-time antinuclear activist whose book, *Energy War: Reports from the Front,* is being published by Lawrence Hill and Co.

Ellyn R. Weiss is a partner in the Washington public interest law firm of Sheldon, Harmon, Roisman, and Weiss and is general counsel to the Union of Concerned Scientists.

Cathy Wolff is a writer in Portsmouth, New Hampshire. She worked on media relations for the Clamshell Alliance from 1976 to 1978.

George R. Zachar writes on nuclear issues from New York City. He formerly edited *Critical Mass Journal.*

Further Information

NUCLEAR ENERGY OVERVIEW

Berger, John. *Nuclear Power: The Unviable Option*. Palo Alto, Calif.: Ramparts Press, 1976.

Caldicott, Helen M. *Nuclear Madness: What You Can Do*. Brookline, Mass.: Autumn Press, 1978. $3.95.

Gyorgy, Anna, et al. *No Nukes: Everyone's Guide to Nuclear Power*. Boston: South End Press, 1978. $8.00.

McPhee, John. *The Curve of Binding Energy*. New York: Ballantine Books, 1975. $1.50.

Nader, Ralph, and Abbotts, John. *The Menace of Atomic Energy*. New York: W. W. Norton, 1977. $4.95.

Nuclear Power, Issues and Choices. Report of the Nuclear Energy Policy Study Group. Cambridge, Mass.: Ballinger Publishing Co., 1977.

Olson, McKinley C. *Unacceptable Risk*. New York: Bantam Books, 1976. $2.25.

Patterson, Walter C. *Nuclear Power*. New York: Penguin, 1976. $3.50.

NUCLEAR TECHNOLOGY, SAFETY, AND HEALTH

Alexander, Peter. *Atomic Radiation and Life*. New York: Penguin, 1965.

Fuller, John. *We Almost Lost Detroit*. New York: Reader's Digest Press, 1975. $8.95.

Gofman, John W., and Tamplin, Arthur. *Poisoned Power: The Case Against Nuclear Power*. Emmaus, Pa.: Rodale Press, 1971. $1.95.

Pizzarello, Donald J., and Witcofski, Richard L. *Basic Radiation Biology*. Philadelphia: Lea and Febiger, 1975.

Sternglass, Ernest J. *Low-Level Radiation*. New York: Ballantine Books, 1972.

Union of Concerned Scientists. *The Nugget File: Excerpts from the Government's Special Internal File on Nuclear Power Plant Accidents and Safety Defects.* U.C.S., 1208 Massachusetts Ave., Cambridge, Mass. 02130. 1979.

———. *The Nuclear Fuel Cycle: A Survey of the Public Health, Environmental, and National Security Effects of Nuclear Power.* Cambridge, Mass.: MIT Press, 1975.

Webb, Richard. *The Accident Hazards of Nuclear Power Plants.* Amherst: University of Massachusetts Press, 1976. $6.95.

NUCLEAR ECONOMICS

Bupp, Irvin C., et al. *Trends in Light Water Reactor Capital Costs in the United States: Causes and Consequences.* Center for Policy Alternatives, Massachusetts Institute of Technology, 1974.

Daneker, Gail, and Grossman, Richard. *Jobs and Energy.* Environmentalists for Full Employment, 1101 Vermont Ave. NW, Washington, D.C. 20005.

Environmental Action Foundation. *Nuclear Power: The Bargain We Can't Afford.* 724 Dupont Circle Building, Washington, D.C. 20036. $3.50.

Grossman, Richard, and Daneker, Gail. *Jobs, Energy and the Economy.* Boston: Alison Publishers, 1979.

Komanoff, Charles. *A Comparison of Nuclear and Coal Costs.* Komanoff Energy Associates, 475 Park Ave. South, New York, N.Y. 10016. $10.

———. *Nuclear Plant Performance Update 2.* (Includes data through December 1977.) Komanoff Energy Associates, 475 Park Ave. South, New York, N.Y. 10016. $5.00.

———. *Power Plant Performance: Nuclear and Coal Capacity Factors and Economics.* Council on Economic Priorities, 84 Fifth Ave., New York, N.Y. 10011, 1976. $5.00.

Lanoue, Ron. *Nuclear Plants: The More They Build, the More You Pay.* Critical Mass, P.O. Box 1538, Washington, D.C. 20013, 1976. $5.00.

Miller, Saunders. *The Economics of Nuclear and Coal Power.* New York: Praeger Publishers, 1976. $15.95.

Taylor, Vince. *Energy: The Easy Path.* Union of Concerned Scientists, 1025 15th St. NW, Washington, D.C. 20005, 1979.

———. *The Easy Path Energy Plan.* Union of Concerned Scientists, 1025 15 St. NW, Washington, D.C. 20005, 1979.

U. S. House of Representatives Committee on Government Operations. *Nuclear Power Costs.* Washington, D.C., April 1978.

NUCLEAR POLICY

Commoner, Barry. *The Politics of Energy.* New York: Alfred A. Knopf, 1979. $4.95.

———. *The Poverty of Power.* New York: Alfred A. Knopf, 1976.

Hayes, Denis. *Rays of Hope: The Transition to a Post-Petroleum World.* New York: W. W. Norton, 1977. $3.95.

Metzger, H. Peter. *The Atomic Establishment.* New York: Simon & Schuster, 1972.

Novick, Sheldon. *The Electric War.* San Francisco: Sierra Club Books, 1976. $11.95.

Warnock, Donna. *Nuclear Power and Civil Liberties: Can We Have Both?* Feminist Resources on Energy and Ecology, P.O. Box 6098, Teall Station, Syracuse, N.Y. 13217, or Citizens' Energy Project, 1413 K St. NW, Washington, D.C. 20005, 1979.

ALTERNATIVES

Clark, Wilson. *Energy for Survival: The Alternative to Extinction.* New York: Doubleday, 1975. $4.95.

Daniels, Farrington. *Direct Use of the Sun's Energy.* New York: Ballantine Books, 1964. $1.95.

Lovins, Amory. *Soft Energy Paths: Toward a Durable Peace.* New York: Harper & Row, 1979. $3.95.

Lovins, Amory, and Price, John. *Non-Nuclear Futures: The Case for an Ethical Energy Strategy.* Cambridge, Mass.: Ballinger, 1975. $5.95.

Lyons, Stephen, ed. *SUN! A Handbook for the Solar Decade.* San Francisco: Friends of the Earth/Solar Action, 1978. $2.95.

Stoner, C. H. *Producing Your Own Power.* Emmaus, Pa.: Rodale Press, 1974. $7.50.

PERIODICALS

Bulletin of Atomic Scientists. 1020 East 58th Street, Chicago, Ill. 60637. Monthly except July and August. $19.50/yr.

Critical Mass Journal. P.O. Box 1538, Washington, D.C. 20013. Monthly. $7.50/yr.

Environmental Action. 1346 Connecticut Ave. NW, Washington, D.C. 20036. Monthly. $15. EA membership fee includes subscription.

Mobilizer. Mobilization for Survival. 3601 Locust Walk, Philadelphia, Pa. 19104.

New Roots: Notes on Appropriate Technology and Community Self-Reliance in the Northeast. U. Mass. P.O. Box 459, Amherst, Mass. 01002. Bimonthly. $8/yr.$6/yr. low income.

Not Man Apart. Friends of the Earth, 124 Spear St., San Francisco, Calif. 94105. Biweekly. $10/yr.

Nuclear Industry, INFO. Atomic Industrial Forum, 7101 Wisconsin Ave., NW, Washington, D.C. 20014. Industry trade association publications. Restricted distribution.

Nuclear News. 244 Ogden Avenue, Hinsdale, Ill. 60521. Official publication of American Nuclear Society.

Nuclear Opponents. Box 285, Allendale, N.J. 07401. Monthly. $3.50/yr.

Nucleonics Week. McGraw-Hill, Inc. 1221 Avenue of the Americas, New York, N.Y. 10021. Industry newsletter. $555/yr.

People & Energy. Institute for Ecological Policies, 1413 K St., NW, Washington, D.C. 20005. Monthly. $10/yr. $8/yr low income.

The Power Line. Environmental Action Foundation, 724 Dupont Circle Building, Washington, D.C. 20036. Monthly, utility news. $15/yr; $7.50/yr low income.

RAIN. Journal of Appropriate Technology. 2270 N.W. Irving, Portland, Ore. 97210. Monthly. $10/yr.

GOVERNMENT SOURCES

Write to these agencies for a list of their publications.

Council on Environmental Quality. 722 Jackson Place, Washington, D.C. 20506.

Department of Energy. Washington, D.C. 20585.

Environmental Protection Agency. 401 M St. SW, Washington, D.C. 20460.

Government Printing Office. Superintendent of Documents, Washington, D.C. 20401.

National Technical Information Service. U.S. Department of Commerce, 5285 Port Royal Rd., Springfield, Va. 22161.

Nuclear Regulatory Commission. Washington, D.C. 20555.

NATIONAL ORGANIZATIONS

These groups can help put you in touch with safe energy activists in your area.

Atomic Industrial Forum. 7101 Wisconsin Ave., NW, Washington, D.C. 20014. 202-654-9260. (Nuclear industry trade association. Pro-nuclear, general information. Good source of statistics.)

Campaign for Political Rights. 201 Massachusetts Ave., NE, Suite 306, Washington, D.C. 20002. 202-547-4705. (Coalition of over 70 organizations. Information clearinghouse and organizing around surveillance and harassment of nuclear power critics. Publications.)

Center for Renewable Resources. 1001 Connecticut Ave., No. 530, NW, Washington, D.C. 20036. 202-466-6880. (Clearinghouse and organizing on promoting solar and other alternative energy sources. Publications.)

Citizens for a Better Environment. 59 E. Van Buren, Suite 1600, Chicago, Ill. 60605. 312-939-1530. (Research on emergency planning and economics, good resource on legal interventions. Publications.)

Committee for Nuclear Responsibility. MPOB Box 11207, San Francisco, Calif. 94101. 415-776-8299. (Scientific and technical research on various nuclear safety and economics issues.)

Critical Mass Energy Project. P.O. Box 1538, Washington, D.C. 20013. 202-546-4790. (Ralph Nader's antinuclear group, specializing in research and lobbying. Publications.)

Edison Electric Institute. 1140 Connecticut Ave. NW, Suite 401, Washington, D.C. 20036. 202-862-3800. (Trade association of the electric utility industry. Good source of information and statistics. Publications.)

Energy Task Force. 156 Fifth Ave., New York, N.Y. 10003. 212-675-1920. (Alternative energy information clearinghouse.)

Environmental Action Foundation. 1346 Connecticut Ave. NW, Washington, D.C. 20036. 202-659-9682. (Specializes in electric utility policies, power demand, and nuclear economics. Publications and information clearinghouse.)

Environmental Action, Inc. 1346 Connecticut Ave. NW, Washington, D.C. 20036. 202-833-1845. (Lobbying, political action, education on solar and other environmental issues. General environment publications.)

Environmental Policy Center. 317 Pennsylvania Ave. SE, Washington, D.C. 20003. 202-547-6500. (Lobbying and research focusing on nuclear waste disposal, power plant siting, health-/safety and conservation.)

Environmentalists for Full Employment. 1101 Vermont Ave. NW, Room 305, Washington, D.C. 20005. 202-347-5590. (Research on employment impact of nuclear and job creation potential of alternative energy sources.)

Friends of the Earth. 124 Spear St., San Francisco, Calif. 94105. 415-495-4770. (Technical analysis of the potential for alternative energy development, research and lobbying on nuclear issues. Publications.)

Institute for Local Self-Reliance. 1717 18th St. NW, Washington, D.C. 20009. 202-232-4108. (Organizing and research on community energy self-reliance. Publications.)

Komanoff Energy Associates. 475 Park Ave., South, New York, N.Y. 10016. 212-686-4191. (Technical research on the economics of nuclear power and coal.)

Mobilization for Survival. 3601 Locust Walk, Philadelphia, Pa. 19104. 215-386-4875. (Coalition of 250 groups organizing around nuclear energy, weapons and disarmament issues. Publications.)

National Clean Energy and Nuclear Energy Moratorium. P.O. Box 1817, Washington, D.C. 20013. 202-547-6661. (Organizing million-signature, safe-energy petition drive by congressional districts.)

National Council of Churches. Energy Project. 475 Riverside

Drive, Room 572, New York, N.Y. 10027. 212-870-2385. (Networking and research on the issues of energy and ethics, the social costs of energy choices, and alternative energy sources.)

Natural Resources Defense Council. 917 15th St. NW, Washington, D.C. 20005. 202-737-5000. (Legal challenges and lobbying. Publications.)

Nuclear Cargo Transportation Project, American Friends Service Committee. Southeastern Regional Office, P.O. Box 2234, High Point, N.C. 27261. 919-882-0109. (Research and organizing around nuclear transport issues.)

Nuclear Information and Resource Service. 1536 16th St. NW, Washington, D.C. 20036. 202-483-0045. (National information clearinghouse for citizens on all aspects of nuclear power. Publications.)

Scientists Institute for Public Information. 355 Lexington Ave., New York, N.Y. 10017. 212-661-9110. (Organization of scientists which periodically issues reports focusing on the economics of energy. General environment publications.)

Sierra Club. 330 Pennsylvania Ave. SE, Washington, D.C. 20003. 202-547-1144. (Large national membership comprising active local chapters. Education and lobbying on nuclear power and alternative energy resources.)

Solar Lobby. 1001 Connecticut Ave. NW, Suite 530, Washington, D.C. 20036. 202-466-6880. (Lobbying on solar and other renewable energy sources.)

Southwest Research and Information Center. P.O. Box 4524, Albuquerque, N.M. 87106. 505-242-4766. (Specializes in nuclear waste and uranium mining issues. Publications.)

Union of Concerned Scientists. 1208 Massachusetts Ave., Cambridge, Mass. 02130. 617-547-5552. (Specializes in technical analysis of nuclear safety problems. Government regulation of the industry. Lobbying. Publications.)

Worldwatch Institute. 1776 Massachusetts Ave. NW, Washington, D.C. 20036. 202-452-1999. (Research on alternatives to nuclear power, such as conservation and solar energy. Publications.)

OTHER ENVIRONMENTAL ACTION FOUNDATION PUBLICATIONS

How to Challenge Your Local Electric Utility: A Citizen's Guide to the Power Industry. 112 pages. Published in March 1974. A basic information and organizing manual for citizens who want to challenge utilities on environmental, consumer, and social grounds. Includes chapters on changing discriminatory rate structures, challenging utility advertising, opposing new power plants and power lines, promoting energy conservation, challenging rate increases, and replacing private utilities with publicly owned systems. $3.50.

Taking Charge: A New Look at Public Power. 100 pages. Discusses the environmental and social benefits of public ownership of electric utilities. Also featured is an analysis of strategies for citizens interested in takeover of the electric companies. $3.50.

A Citizen's Guide to Electric Rate Structures. (Tentative title. Forthcoming.) A technical and organizing manual for citizens seeking reform of promotional utility-rate structures and price hikes.

A Citizen's Guide to the Fuel Adjustment Clause. 52 pages. Describes utility abuses of the fuel adjustment clause (which accounted for 65 percent of the increased cost of electricity in 1974) and outlines strategies and methods to challenge these abuses. $2.50.

Utility Scoreboard. Compares the nation's 100 largest power companies on 15 separate environmental and consumer issues such as rate structures, air and water pollution controls, tax overcharges, advertising expenditures, and excess generating capacity. $3.50.

Phantom Taxes in Your Electric Bill. 28 pages. Explains how utilities charge their customers for billions in federal income taxes that are never paid. Lists overcharges by each of the nation's 150 largest power companies and explains what citizens' groups can do about it. $2.50.

The End of the Road: A Citizen's Guide to Transportation Problemsolving. 160 pages. Explains the financial and political aspects of transportation planning on the local level and outlines how to use effectively the Department of Transportation, the National Environmental Policy Act, the Clean Air Act, and other regulations. $3.50.

Resource Recovery: Truth and Consequences. 80 pages. Describes resource recovery processes and problems. Contrasts the economics and energy costs of high- and low-technology waste-handling systems.

The Environmental Action Foundation is a nonprofit, public interest organization engaged in research and education. Its goal is to have citizens understand and participate actively in complex environmental debates. An outgrowth of Earth Day, 1970, EAF integrates environmental concerns with resource and energy shortages, economics, and consumer injustices.

Nuclear power is only one of the Foundation's several projects. The staff also investigates electric utilities, solid and hazardous waste, and materials conservation.

Environmental Action Foundation
724 Dupont Circle Building
Washington, D.C. 20036
202-659-9682

Environmental Action Foundation staff:

Claudia Comins, Kathy Durso-Hughes, Andrew Feeney, Lois Florence, Casey Grant, Anne Woiwode, Alden Meyer, Richard Morgan, Helen Sandalls, Liz Waldo, and Annette Woolson.

Index